Venerable Trees

Venerable Trees

History, Biology, and Conservation in the Bluegrass

Tom Kimmerer

UNIVERSITY PRESS OF KENTUCKY

Copyright © 2015 by Tom Kimmerer

Published by The University Press of Kentucky,
scholarly publisher for the Commonwealth,
serving Bellarmine University, Berea College, Centre College of Kentucky, Eastern Kentucky University, The Filson Historical Society, Georgetown College, Kentucky Historical Society, Kentucky State University, Morehead State University, Murray State University, Northern Kentucky University, Transylvania University, University of Kentucky, University of Louisville, and Western Kentucky University.
All rights reserved.

Editorial and Sales Offices: The University Press of Kentucky
663 South Limestone Street, Lexington, Kentucky 40508-4008
www.kentuckypress.com

Maps created by the author using public-domain data. Basemaps used by permission, © 2014 ESRT, DeLorme, NAVTEQ. All rights reserved. Unless otherwise noted, photographs are by the author.

Cataloging-in-Publication data is available from the Library of Congress.

ISBN 978-0-8131-6566-0 (hardcover : alk. paper)
ISBN 978-0-8131-6567-7 (epub)
ISBN 978-0-8131-6568-4 (pdf)

This book is printed on acid-free paper meeting
the requirements of the American National Standard
for Permanence in Paper for Printed Library Materials.

Manufactured in the United States of America.

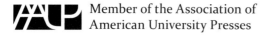

Member of the Association of
American University Presses

The financial support of the following is gratefully acknowledged:

VENERABLE TREES, INC.

The mission of Venerable Trees, Inc., is to conserve the ancient trees of the Kentucky Bluegrass through outreach, publications, education, and research, and to promote the planting of native trees in the urban and agricultural landscapes of the Bluegrass.

BALL HOMES AND DON AND MIRA BALL

Ball Homes is a leader in the housing market in Central Kentucky. In 2014, the company received an award from the Lexington–Fayette County Environmental Commission for its ongoing efforts to preserve ancient trees in the Bluegrass.

To Stacey, Will, Greg, and Allison Kimmerer,
Linden, Andy, and Dylan Lane,
Larkin and Dave Podsiedlik,
with love

Contents

Introduction: Two Trees 1

1. The St. Joe Oak: Finding Venerable Trees 5
2. The Woodford Groves: The Bluegrass Today 17
3. The Llama Tree: Venerable Trees in the Nashville Basin 35
4. Venerable Tree Species 47
5. The Ingleside Oak: The Bluegrass and the Nashville Basin in 1779 71
6. The Woodland Pasture 87
7. The Mother Tree: Reproduction of Venerable Trees 99
8. The Guardians: Trees in Cemeteries 113
9. The Loudon Grove: Trees in Public Spaces 121
10. The Coldstream Tree: Groves, Remnants, and Developments 139
11. The Elmwood Trees: Growing Old Gracefully 151
12. The Floracliff Trees: The Long Lives of Venerable Trees 165
13. The Old Schoolhouse Oak: Extending the Lives of Old Trees 177
14. The Future of Venerable Trees and Woodland Pastures 189

Acknowledgments 207

Appendix: Trees of the Bluegrass and Woodland Pastures 211

Notes 217

Index 225

Color photographs follow page 118

Figure I.1. A bodhi tree in Sumatra, Indonesia.

Figure I.2. A bur oak in Lexington Cemetery.

Introduction

Two Trees

Two trees stand on opposite sides of the Earth. One is a bodhi tree near the grave of the royal family in the Karo Highlands of Sumatra. The tree is guarded by, or perhaps is guarding, two stone men, covered in moss and lichen, that nearly blend into the tree. The other tree is an old bur oak shading a grave in the Bluegrass of Kentucky. The headstone has eroded with age and, like the stone men, is draped in moss and lichen. Both trees are ancient, with the stout cylindrical stems and coarse branching of very old trees.[1]

People in all cultures revere large, old trees. From temple trees in Southeast Asia to the giant sequoia and coast redwood groves of the American West, people visit, worship, and love trees. Trees are venerated—held in awe and esteem.

Our relationship with trees is both practical and reverential. We can easily appreciate a large, old oak tree while comfortably dining at a beautiful oak table. We have always used trees for food, fiber, fuel, and shade. But our relationship with trees also has a deeply spiritual context that dates back to the time before we were even human. Our ancestors were arboreal creatures living in and among trees. When we left the forest, we did not leave the trees behind.

Most of the world's population is now urban, yet even in cities we surround ourselves with trees. Urban trees have practical utility, cooling buildings and sidewalks, cleaning the air, providing shade. But it is the spiritual connection we maintain with trees that compelled us to bring them along when we left our ancestral forests.

This book is a celebration of the long relationship between people and trees and a cautionary tale of what happens when we neglect that relationship. I will focus on very old trees in two closely related urban and agricultural areas, the Bluegrass of Kentucky and the Nashville Basin. Most of these trees were present before the first permanent settlement in the late 1700s and remain with us today, but they are disappearing and not being replaced. I will refer to these as venerable trees to indicate their great age and value, though with only a few exceptions their exact ages are unknown.

My first venerable tree, the first one that I spent time with and cared about, was an old American beech, rotten at its base, offering plenty of room for a fort, cave, or superhero's lair. My beech was a treasure trove for boys and girls in my Baltimore neighborhood; fungi popped from its bark or roots in spring and fall, and insects, fence lizards, snakes, and raccoons wandered in and out of the crevices in the rotting stem. We relished the sweet crunchy beechnuts in the fall and enjoyed the work of extracting them from their shells. There were oaks, maples, plenty of trees whose names I did not know until a decade later. But it was the beech that was the center of our woodland play, and the beech that I remember individually among all the trees of my childhood. Eventually it died and fell, but I can still find traces of the old tree as an umber stripe of decayed wood under leaf litter. Years later, I became a forest scientist and have spent many years devoted to the practical utility of trees. But I never completely lost the reverence for that old beech and all the subsequent trees in my life.

In 1982 I came to Kentucky and soon moved with my family to a small farm in Garrard County, thirty miles south of Lexington. Each day on my commute, I would see giant trees on farms, in abandoned pastures, and in industrial areas. Many of them had the dead tops, or stag-heads, that indicate great age, decline, lightning strikes, or all three. The oldest trees that I saw were bur oak, blue ash, Shumard oak, chinkapin oak, and kingnut, and I began thinking of them as the venerable trees of the Bluegrass.

I began keeping track of all the venerable trees I saw. Over the ensuing years, many of these trees have become very familiar to me, and I watch them change over the seasons and years. Some have died catastrophic deaths, taken by lightning or bulldozers. Others died more slowly for more subtle reasons, and some of these remained standing for decades after their deaths. Still others, but ever fewer, remain hale and hearty, shading horse pastures and gas stations alike.

I have also sought out young trees of these venerable species, usually in vain. One day in the late 1980s, while scouting locations for teaching field classes, I chanced upon an elderly blue ash on the edge of a field. When I walked out into the field to see this giant, I realized that the hedgerow I was walking along had blue ash saplings and seedlings in abundance. There were hundreds of young blue ash trees, all apparently the progeny of the huge mother tree. I paid scant attention to the signs of impending doom—the survey stakes, the distant bulldozer. A few days later, I brought my class to the site. The mother tree remained, but all her progeny had been bulldozed into piles, slowly burning in the morning breeze. A sign indicated a new housing development, Ashbrook. Even the venerable mother tree is gone now, having succumbed to soil compaction and the insults of suburban lawn care.

We badly need vigorous efforts to extend the lifetime of our ancient trees through better public engagement, better management, and better policies. We also need to plant more of the most characteristic native species from local seed sources and manage them properly. We need to take action to ensure that the venerable trees of our current landscape are sustained through all the generations to come.

Beyond these practical considerations, we must be more mindful of the presence of venerable trees in our midst. We may travel thousands of miles to see famous old trees like the redwoods, while ignoring trees equally worthy of our veneration in the landscapes in which we live.

I wrote this book not to recount how much we know about these trees, their management, and their future, but to remind us how little we know. The more people become aware of the ancient trees in our landscapes, the more we will be stimulated to see them clearly, tend them carefully, conserve them vigorously, and try to understand them fully. Most important of all is my hope that more of us will venerate these trees enough to ensure their future.

1

The St. Joe Oak

Finding Venerable Trees

The huge bur oak tree stands surrounded by concrete and cars in the middle of a two-story parking structure at the St. Joseph Medical Center in Lexington, Kentucky. The tree has all the hallmarks of a very old tree. It is massive. The branches are gnarled, the leaves tufted on the ends. The stem bears scars of old wounds and lost branches. The heavy cylindrical stem passes downward through two levels of the parking structure, then through a metal grate into a concrete enclosure. There are many other trees nearby, but no other ancients like this one. I'll call this tree the St. Joe Oak.

The medical complex was not here sixty years ago. Instead, there was a pasture shaded by large, old trees, clearly visible in a 1955 aerial photograph. This woodland pasture included about twenty-five old bur oaks (*Quercus macrocarpa*), along with many chinkapin oaks (*Quercus muehlenbergii*), Shumard oaks (*Quercus shumardii*), blue ash (*Fraxinus quadrangulata*), and kingnut hickories (*Carya laciniosa*).

The woodland pasture is gone now, and only the St. Joe Oak still stands. On the basis of its size and form and historical records, we can be confident that this tree was present on this spot not just before the parking structure was built, but before Lexington, or Kentucky, or the United States existed.

St. Joseph Hospital planned to cut the tree down to build a parking structure in 1990. Local citizens, especially those who lived and worked in the neighborhood and valued the old tree, insisted that it be preserved.

Figure 1.1. The St. Joe Oak.

Figure 1.2. A 1955 aerial photo of the St. Joe Oak's location on Harrodsburg Road in Lexington, indicated by the arrow. Campbell House Hotel is to the right. The woodland in the foreground is the present location of St. Joseph Hospital. Photograph by Lexington-Fayette Urban County Government Division of Planning.

Donna Westphal, who worked nearby, said, "There's something spiritual about a tree that old, a creature that's been on Earth for centuries." She and others waged a campaign to save the tree, and they eventually convinced the hospital administrators to save the tree.[1]

St. Joseph Hospital built the parking structure around the tree in the early 1990s, providing the tree with soil volume beneath the structure and avoiding injury and soil compaction during construction. Over time it has become clear that the soil volume is inadequate and the tree is not receiving proper care. It is in decline and will die unless further steps are taken to protect it.

Despite its isolation, this tree is not alone. There are groves and individual trees that predate European settlement throughout both the Bluegrass of Kentucky and the Inner Nashville Basin of Tennessee and that number in the thousands. Extensive woodland pastures covering many acres are common on farms in the Inner Bluegrass. In the Nashville Basin, there are many individual venerable trees, but only a few extensive woodland pastures.

In more developed areas of the Bluegrass and Nashville Basin, there are only a few remaining groves of multiple trees, but there are a large number of individual ancient trees that, like the St. Joe Oak, remained behind when the woodland pastures were developed.

The Bluegrass is famous for its horse farms, with their gently rolling

Figure 1.3. Venerable trees in a rural woodland pasture in the Inner Bluegrass.

Figure 1.4. Venerable trees in a rural woodland pasture in the Inner Nashville Basin. Intact woodland pastures are less common in the Inner Nashville Basin than in the Inner Bluegrass.

Figure 1.5. A grove of venerable trees in an urban park in the Inner Bluegrass.

green pastures, distinguished houses, plank fences, and barns. Without the huge old trees in the pastures and around the houses, the Bluegrass would look barren and uninteresting. Yet the old trees are disappearing at an alarming rate. Many things contribute to their loss, including old age. The old trees are being replaced, but not with species that will live as long as the original trees.

To understand the origin of these trees, why they are still here, and why they are disappearing, we have to go back to the founding of Lexington. The town was founded in June 1775 and named for the famous battles at Lexington and Concord two months earlier, but no permanent structures were built until Robert Patterson was sent from Harrodsburg in 1779 to establish a settlement.

Patterson brought with him a party of men, including a young Virginian, Josiah Collins. Many years later, Collins recounted his experiences as one of the people who built Lexington. "I was at the beginning of that

Figure 1.6. A brass plaque of the blockhouse built in 1779 by Josiah Collins and others. The plaque is at the corner of Mill and Vine streets in Lexington, the approximate original location of the blockhouse.

place," Collins recalled of April 1779. He was almost twenty-two and had come from Virginia to Fort Boonesborough via the Wilderness Road and then to Fort Harrod, now Harrodsburg.

On April 17, Josiah Collins used his ax to fell a large bur oak. From this tree he built a blockhouse, a simple fortress for protection from Indians and the first permanent structure in Lexington, which was then part of Virginia. He later recalled that "when there was a town built there and he an old man, he could say he had fallen the first tree cut on the spot." Asa Farrar, another early settler, felled another bur oak to clear what is now Main Street, and he said it was too large to be cut with any saw.[2]

The blockhouse no longer exists, but one of the first permanent cabins still stands. The Patterson Cabin was built of oak and walnut at about the same time as the blockhouse, probably by Josiah Collins. Patterson used the cabin as his residence. Once he was wealthy enough to build a larger house, he used the cabin as slave quarters and later as a tool shed. He must have had a fondness for the cabin, for he took it with him when he moved to Dayton, Ohio, in the early 1800s. The cabin eventually returned to Lexington, where it now resides on the campus of Transylvania University. We can guess, but do not know for certain, that some of the wood in that cabin came from the venerable oaks that were felled by Collins and Farrar and the other hundred or so early settlers.

Collins had his choice of big trees—there was an abundance of large bur oaks, chinkapin oaks, Shumard oaks, blue ash, and kingnuts. There were also American elms (*Ulmus americana*), hackberries (*Celtis occidentalis*), sugar maples (*Acer saccharum*), black maples (*Acer nigrum*), white ash (*Fraxinus americana*), and black walnuts (*Juglans nigra*). But it was bur oak, chinkapin oak, Shumard oak, blue ash, and kingnut that characterized the Kentucky Bluegrass when European settlers first arrived and that are still with us today.

Today Lexington is a thriving city of 300,000 ringed by some of the most valuable farmland in the world. Yet many trees that were here when Josiah Collins first hefted his ax are still here. The rest of the Bluegrass and the Nashville Basin was settled at about the same time. Horse and cattle farms, towns, cities, and suburbs have replaced the land that Collins saw, but the trees are still with us. These venerable trees, two hundred to five hundred years old, or perhaps older, survived the conversion of native woodland pastures to farms and have persisted even with the development of modern cities.

Figure 1.7. The Patterson Cabin, built in 1779 or 1780, now stands on the grounds of Transylvania University in Lexington. The cabin was built of bur oak and walnut and had a stone-and-stick fireplace and chimney.

Figure 1.8. A detail of the Patterson Cabin showing an old oak log. Most of the original wood has been replaced, but this may be one of the original bur oak pieces used to build the cabin.

Figure 1.9. A lone ancient bur oak in an industrial area, the remnant of a woodland pasture.

Now, with the growth of the city, the old trees are found in unlikely places—behind gas stations, in used-car lots, next to a motel. Many are in the middle of pastures, shading horses and cattle. These trees are huge, their low, thick branches evidence that they grew in the open, not in the shade of a forest.

Slowly, with time, development, and poor management, the ancient trees are declining in numbers and in health. Old trees grow slowly and die slowly. I have observed many of these trees for more than thirty years, and some have been in slow decline for all that time. How much longer will we have these venerable trees, and how many are left?

For some answers, we can turn to Ursula Davidson, a young graduate student at the University of Kentucky in 1950. She conducted a census of old bur oaks in the Bluegrass as part of her dissertation research.

Davidson traveled all the roads of Fayette County by car, carefully counting and assessing all the bur oak trees she could see from the road or in parks and other public spaces. She chose bur oak because she could easily recognize the ancient trees, and because the old blue ash trees were too numerous to count easily. She was not able to count trees on private lands except for those visible from the road, and there were many trees deep in the large horse farms that she could not see. Lexington in 1950 was a small city of 100,000; horse farms occupied most of the county, and the farms were not open to the public. A survey by car was the best method available to her at the time.

There was another reason that she counted trees from a car. As a child, she had contracted polio and was unable to walk. At a time when few people with disabilities were able to attend college, Davidson earned her undergraduate degree from Morehead State University and her master's in botany from the University of Kentucky. She went on to a long and distinguished career teaching English and science in eastern Kentucky.

Davidson counted and mapped 370 bur oaks in Fayette County, including the St. Joe Oak and its companions. Many of the trees she found were in groves of ten to fifty bur oak trees, along with larger numbers of blue ash and a few chinkapin oaks, Shumard oaks, kingnuts, and other species. Many of these groves were in horse farms that today are housing developments, hospitals, churches, and parking lots.[3]

Mary Wharton, a botanist at Georgetown College, traveled the same roads in 1978 and found 178 bur oaks. She predicted that all but about thirty would be gone by 2000.[4]

In 2013, a 1950 AAA road map of the Bluegrass in hand, I followed the routes that Davidson and Wharton had explored, counting all the bur oaks I could see from the road. To sample in the same way as Davidson, I did not get out of my car. Some of the roads have changed, replaced by housing developments or newer roads, but I was still able to follow her route. I found 43 bur oaks visible along those roads, compared with Davidson's 370. Few of the remaining bur oaks were in groves, but were scattered individuals, alone or together with blue ash trees. I was later able to get a closer look at most of the bur oaks on foot.

There are many more roads in the county today, as development has enlarged the city. Many of these roads punched through the horse farms that were inaccessible to Davidson. On those roads, I found another 46 bur oaks. In city parks and schoolyards, there are 18 more. That brings the total to 107, not counting trees deep in the horse farms. There are another 103 bur oaks in Lexington Cemetery, about a quarter of which are ancient trees.

The bur oaks of Fayette County have been reduced in number by nearly 90 percent in sixty three years, and many of the remaining trees are in poor shape. Most striking is that almost all the bur oaks in Fayette County today are individual trees, not growing in the extensive groves that Davidson found. Other counties in the Bluegrass and Nashville Basin currently have more trees and are less highly developed, but we have no census data like Davidson's for comparison. Blue ash trees have fared better. There are still substantial groves of blue ash on farms and scattered throughout our urban forest. Many of these venerable trees are in decline, dying or dead.

Where did these venerable trees come from and how did they survive development? The combination of bur oak, chinkapin oak, Shumard oak, blue ash, and kingnut is uncommon. Shumard oak is a southern species, primarily of the Atlantic and Gulf Coastal Plain and Mississippi River valley. Bur oak, chinkapin oak, and blue ash are upland midwestern trees of sweet (neutral or alkaline) soils, especially on limestone uplands. Kingnut is common along streams and on limestone uplands. As a group, these species are abundant only in the Bluegrass and the Nashville Basin, both formed from the limestone rock of the Cincinnati Arch.

How old are these trees? In most cases we do not know. Assessing the age of trees is laborious and requires special training. Some species grow faster than others, and trees grow faster on better sites. A pin oak (*Quercus palustris*) four feet in diameter in an urban setting may be less than eighty years old, whereas a two-hundred-year-old blackjack oak (*Quercus marilandica*) on top of a dry ridge may be only four inches in diameter. As we shall see later, there is little relationship between the size of a tree and its age. Some of the oldest trees in the world are small, having grown in very restricted conditions. Many giant trees are quite young, including most of the large oaks on streets and in yards and parks.

In 1980 Bill Bryant and his colleagues estimated the age of trees in a remnant Bluegrass woodland pasture. Their results were remarkable. Walnuts were less than 150 years, but many bur oaks, Shumard oaks, chinkapin oaks, and blue ash were over 400 years of age, and other trees were more than 200 years old. These trees were clearly present in the Bluegrass before the first European settlers and African slaves arrived here.[5]

The mystery about these trees is not why they are disappearing, but why they are still here. In almost all urban areas of North America, trees present at the time of first settlement did not survive. Ax and crosscut saw felled nearly all of them, for houses, fences, and fuel, but most of all to get them out of the way, to make room for the new settlers with their buildings, roads, and farms.

A clue to their survival may lie in the observation that these trees appear open-grown, without the tall, straight trunks of forest trees. A

Figure 1.10. A remnant woodland pasture in an urban park. This stand includes some native giant cane.

further clue is that they are not tolerant of deep shade and tend to grow well only in the open, at least when mature. Early accounts of the region and the trees themselves tell us that before European settlement, there were areas of the Bluegrass and the Nashville Basin consisting of open-grown trees, widely spaced and shading a grassy or herbaceous understory. These are woodland pastures.

A woodland pasture, as I use the term in this book, is an area with very large trees with broad, spreading crowns, which indicate that they did not grow up in a forest but were widely separated. Underneath the large trees are grasses and herbs. The original ground cover consisted of native grasses and herbs and giant cane.[6]

The first farmers quickly replaced the native grasses and cane with forage crops, but most of the trees were kept. The trees would have been more important as shade trees than as timber trees. The Bluegrass and the Nashville Basin provided fine grazing without the labor of extensive land clearance, and livestock would have benefited from shade.

Soon after the initial farms were established, the great horse farms, with their characteristic mansions, fences, trees, and cattle were created.

The venerable trees were a natural and valued feature, providing character to the mansions' landscapes and shade for the prized horses.

Today, in spite of the steep decline in numbers of venerable trees, there are still groves of these old giants throughout the Bluegrass and the Inner Nashville Basin. To understand why they are here, we need to look at what makes the Bluegrass and Nashville Basin so unusual, and consider the origin of the ancient groves.

2

The Woodford Groves

The Bluegrass Today

Woodford County, Kentucky, is the only county entirely within the Inner Bluegrass. The mostly rural county retains a very large number of woodland pastures spread across many horse and cattle farms. Many of these farms have not changed much since the late eighteenth century, when they were founded. Some are still held by the founding families. To the first-time visitor, a tour of Woodford County is a source of astonishment.

It is the scenery that strikes the visitor first—the gently rolling pastures with black or white plank fences, elegant estate houses and barns, and woodland pastures of huge, old trees. Anybody who visits the Bluegrass is struck by how unusual this landscape is, resembling no place else on Earth.

The Bluegrass is an ecological region defined by its gently rolling hills on limestone bedrock. The name of the region apparently derives from the bluish color of the seed heads of unmowed grass in summer, though some people perceive a bluish cast to the grass in the early spring. Kentucky bluegrass (*Poa pratensis*) is originally from Europe and western Asia, not Kentucky, however. Originally called meadow grass, bluegrass was brought to Kentucky by early settlers, and it quickly replaced the native grasses and giant cane (*Arundinaria gigantea*). Kentucky bluegrass was planted so widely and so early that biologists don't know with certainty which native grasses may have been here before 1779. Kentucky is often called the Bluegrass State, though the Bluegrass Region occupies only a quarter of the state. The

Figure 2.1. This woodland pasture in Fayette and Woodford counties covers multiple farms and several thousand acres.

Figure 2.2. A Woodford County development. This pasture is part of a rural housing development that has concentrated houses and kept the woodland pasture intact.

Figure 2.3. An old chinkapin oak on a Bluegrass farm. Many farms have lost their original woodland pastures, but individual trees remain.

term *bluegrass* is applied to the region, the state, the grass, and the music, although neither the grass nor the music originated in the Bluegrass Region.

The Bluegrass Region is divided into three distinct subregions, the Inner Bluegrass, the Hills of the Bluegrass (also called the Eden Shales), and the Outer Bluegrass. The Inner Bluegrass consists almost entirely of limestone, mostly of a single formation called the Lexington Limestone. The Lexington Limestone is hundreds of feet thick and sits on top of another limestone layer. Together the limestone layers are up to 1,400 feet thick. These rocks are ancient, dating from the Ordovician Period, about 450 million years ago. The Inner Bluegrass is highly fertile, its deep, well-drained soils covering the limestone bedrock. The high phosphorus content of the Lexington Limestone creates nutritious forage that forms the basis for our famous Thoroughbred horse industry.

The Outer Bluegrass is somewhat less fertile than the Inner Bluegrass, and the Hills of the Bluegrass are less fertile still. Most of the existing woodland pastures of venerable trees are today found in the Inner Bluegrass; scattered smaller stands and individual trees can be seen in the Outer Bluegrass, but there are very few in the Hills of the Bluegrass. Farmers largely abandoned the Hills of the Bluegrass by the middle of the twentieth century

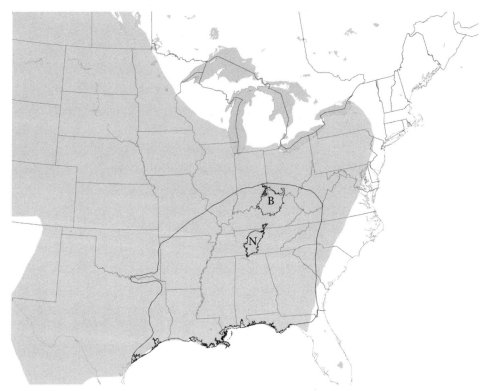

Figure 2.4. Map of the eastern United States, highlighting the Bluegrass (B) and Nashville Basin (N). The Bluegrass is largely in Kentucky; portions of the Outer Bluegrass extend into Ohio and Indiana. The Nashville Basin is largely in Tennessee, though there is a small extension of the Outer Nashville Basin into Kentucky to the north and Alabama to the south. The map also shows the approximate ranges of American bison (gray) and giant cane (solid line) in the period just before European settlement of North America.

because of the low fertility of its soils, and the farms have largely been taken over by scrubby forests of eastern redcedar (*Juniperus virginiana*) and hardwoods.

The Bluegrass is surrounded on three sides by the Knobs Region, a chain of steep hills capped in hard limestone or sandstone underlain by softer shale. The Knobs bring an abrupt end to the vegetation of the Bluegrass, replacing the limestone-loving species with trees more characteristic of the acidic soils of the Appalachian Mountains.

Although the Bluegrass is fertile and gets adequate rainfall, it is very susceptible to drought. The limestone bedrock creates a landscape known as karst. Karst is a German word originally used to describe a limestone region in Slovenia, but today it is applied to regions throughout the world where highly fractured limestone forms the primary bedrock. In karst landscapes,

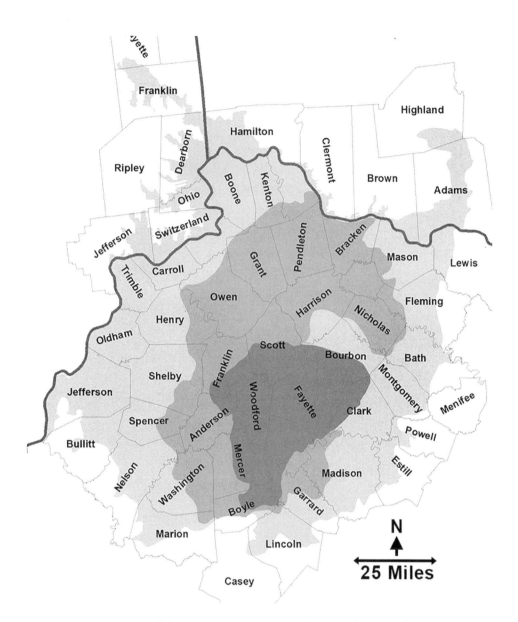

Figure 2.5. Ecoregions of the Bluegrass. The Inner Bluegrass (dark gray) is the most fertile area and includes the largest remaining woodland pastures. The Hills of the Bluegrass (medium gray) is the least fertile region and today consists largely of small farms and young forests of hardwoods and cedar; it contains no old woodland pastures. The Outer Bluegrass (light gray) is similar to the Inner Bluegrass, but it has more rolling hills and greater variability in soil quality. Counties that are at least partly within one or more Bluegrass ecoregions are labeled. Only Woodford County is entirely in the Inner Bluegrass.

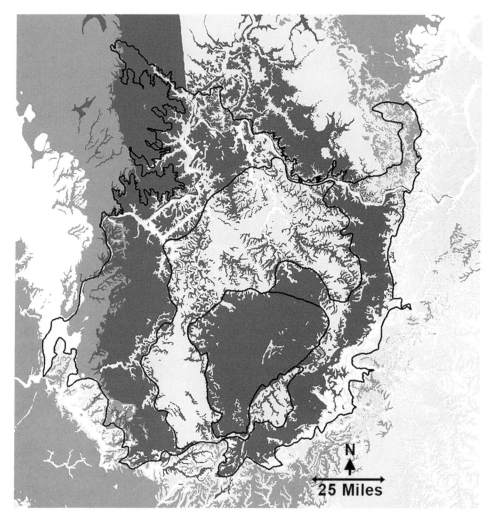

Figure 2.6. Karst (gray) in the Bluegrass. Almost the entire Bluegrass is karst topography on limestone. The Inner and Outer Bluegrass are almost entirely on limestone. The Hills of the Bluegrass have beds of mixed limestone and shale.

rainfall slowly dissolves the rock, creating cracks, fissures, sinkholes, and caves. Rainfall can move rapidly through the soil into the limestone rock and down into groundwater. Even a moderate deficit of rainfall turns the land brown and dries up the creeks.

For most plants, frequent droughts are a huge challenge. But the trees of the Bluegrass have a big advantage over smaller plants: they can root deeply into the cracked rock and get water and nutrients from very deep, sometimes hundreds of feet below ground. It is fairly common to find trees in the Bluegrass that have few or no surface roots but are growing very well.

One of my favorite trees is a white ash at the center of a restaurant patio near my home. I have enjoyed sitting under this tree for over thirty years. The tree is healthy, vigorous, and growing well, in spite of the fact that it has no surface roots or access to soil and is surrounded by concrete. In many locations, such a tree would be doomed, but not in karst: this tree is deeply rooted in the Lexington Limestone, and people will probably be sitting under it for many years to come.

Today the Inner Bluegrass is a mix of livestock farms, including our famous horse farms, and urban areas. Although soils are fertile, drought restricts the growth of row crops, and the dominant land use is pasture.

Urbanization is the greatest threat to the continued existence of our venerable trees in both the Bluegrass and the Nashville Basin. In the Bluegrass, Lexington has grown to a city of over 300,000 people. Several organizations, including the Bluegrass Conservancy and the Fayette Alliance, have been formed to slow the spread of urban areas, and they have been fairly successful. In 1974 the city of Lexington and the county of Fayette were merged into a single urban county government. The unified Lexington-Fayette Urban County Government was able to impose much stronger land use restrictions on the rural parts of the county than had previously been possible. The merged government delineated an urban boundary within which to concentrate development, and this has relieved some of the pressure for development of farmlands in the rural parts of the county. Today the famous horse farms form a ring around the city of Lexington. Fayette County has adopted strict land-use regulations and a purchase-of-development-rights (PDR) program designed to concentrate development within the current urban boundary and allow the valuable farms to continue operating. In the publicly funded PDR program, landowners sell conservation easements for their land but continue to operate the farms as they always have. This provides an opportunity to preserve woodland pastures on these farms without the threat of development. To date, the PDR program has protected more than 28,000 acres of land in the rural service area of Fayette County; the goal is eventually to protect 50,000 acres from development. While the purpose of the PDR program has been to preserve operating farmland, it has also preserved a substantial amount of woodland pasture on those farms.

The Bluegrass Conservancy is a privately funded land trust that purchases conservation easements on farms throughout the Inner Bluegrass and has permanently protected more than 20,000 acres from development. The Fayette Alliance is a conservation organization that advocates for smart

Figure 2.7. A white ash deeply rooted in karst. This tree has no access to soil and no surface roots. It is growing well despite its tight quarters because its roots are deep in the fractured limestone.

Figure 2.8. A clear view of the tree pictured in figure 2.7.

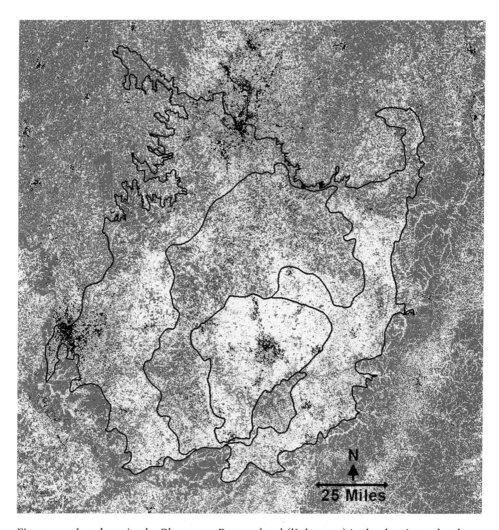

Figure 2.9. Land use in the Bluegrass. Pastureland (light gray) is the dominant land use in the Inner Bluegrass and the parts of the Outer Bluegrass where the primary rock is limestone. Forests (medium gray) are predominant in the Hills of the Bluegrass and along river corridors. Urban areas (dark gray) occupy about one-quarter of the land area of the Inner and Outer Bluegrass.

land use and for keeping urban development within the urban service boundary in Fayette County.

The result of the efforts of these governmental and conservation organizations has been the establishment of some of the strictest land-use regulations in the United States, which have limited the loss of woodland pastures to development. Lexington has continued to grow within its urban boundary. In spite of the growth of the city's population, there is still adequate land for development within the city, and the Urban County Council has recently renewed its commitment to maintaining the current urban boundary.

Some counties have been less successful at slowing development. In Woodford County, the only county entirely in the Inner Bluegrass, strict land-use regulations at the northern end of the county have protected most of the existing woodland pastures, but the southern half of the county has been gradually broken up into smaller and smaller landholdings, and some of these have become housing developments.

Land-use planning in other counties in the Bluegrass has had mixed success. Many counties have few or no land-use restrictions that prevent the conversion of farms and woodland pastures to housing and commercial real estate development.

There is a direct relationship between woodland pastures and horse farms. In the Inner Bluegrass, extensive woodland pastures are found almost entirely on horse farms, or on farms with a mix of horses and cattle. This is true especially in Fayette, Woodford, Bourbon, and Scott counties, the counties with the largest number of woodland pastures and the largest number of horse farms.

In the Outer Bluegrass, we can find a large number of individual venerable trees or small groups of trees, but we only occasionally find extensive woodland pastures like those in the Inner Bluegrass. We don't know whether there were extensive woodland pastures in the Outer Bluegrass at the time of settlement. It is possible that woodland pastures were less common in the Outer Bluegrass than in the Inner Bluegrass before settlement. It is also possible that differences in land use caused the loss of woodland pastures in the Outer Bluegrass. Horse farms are less common in the Outer Bluegrass, where there is a greater emphasis on cattle, hay, and row crops. It may be that woodland pastures were cleared on these farms. We probably will never know.

ALONG THE RIVERS

The Kentucky River meanders through the Bluegrass, cutting deeply into the limestone and creating steep bluffs known as the Palisades. Forests line both sides of the river, and the land there is too steep for farming, except on a few narrow river bottoms. The forests of the Palisades are rich, fragrant, and lush. They are an interesting blend of Appalachian and midwestern tree species. Yellow buckeye (*Aesculus flava*), an Appalachian tree, grows right beside the midwestern Ohio buckeye (*Aesculus glabra*), and hybrids of the

Figure 2.10 Bluffs and bottomland forests along the Kentucky River. Tree species of the woodland pastures are common on the bluff tops and on benches and slopes on the bluff faces.

two species are common. Yellowwood (*Cladrastis kentukea*), a lovely, small tree with spectacular white flowers, is common here but rare elsewhere. At one time, huge eastern redcedars lined the river on the bluffs and upper slopes, but redcedars are so useful for insect-resistant furniture and closets that the big trees are all gone.[1]

The Licking and Salt rivers are the other two major rivers in the Bluegrass. They cut less deeply into the limestone. As a result, the ability to farm right up to their edges has altered the forests more than is the case along the Kentucky River.

The trees of our woodland pastures can be found along the rivers. Bur oak is occasionally found along the Kentucky River bluffs, while chinkapin oak, blue ash, and kingnut are common.

Shumard oak is abundant throughout the Bluegrass in wooded slopes, bluffs, and woods along creeks. Many other red oaks can be found in the same woods, creating a confusing mélange of species and hybrids. Northern red oak (*Quercus rubra*), pin oak (*Quercus palustris*), scarlet oak (*Quercus coccinea*), shingle oak (*Quercus imbricaria*), and southern red oak (*Quercus falcata*) are red oaks found in at least some Bluegrass counties. In the

Figure 2.11. Creeks near the Kentucky River are forested, which provides nurseries for venerable tree species, especially in nature preserves.

woods along the slopes and bluffs above the rivers, and along major creeks, it is very common to find trees that appear to be intermediate between two of these species. Some of them can be recognized as a particular hybrid, such as riparian oak (*Quercus × riparia*), a hybrid of Northern red oak and Shumard oak, which I have seen many times along Bluegrass waterways. The majority of these intermediate oaks cannot be recognized for certain, meaning that their parentage is not obvious. It is possible that the red oaks of the Bluegrass form a hybrid swarm, a complex mixture of species that interbreed to form hybrids of mixed parentage.

As we will see (chapter 5), woodland pastures are no longer regenerating themselves. While there are many very old trees, and a few middle-aged trees, there are almost no young trees. Constant grazing by livestock and frequent mowing prevent the pasture trees from reproducing. The forests along the Kentucky, Salt, and Licking rivers and some of their tributary creeks are nearly the only places where the trees of the woodland pastures are reproducing naturally. Most of the young blue ash trees that I have seen in recent years have been in the forests above the Kentucky River and the larger creeks that feed it. Young chinkapin and Shumard oaks are common on the upper slopes of the larger rivers. Kingnut is found both on the tops of the bluffs and along the river bottoms.

The forests of the Kentucky, Licking, and Salt rivers and their larger tributaries provide a nursery and refuge for the trees of the woodland pastures. Development has been a constant threat to these forests, especially along the tops of the bluffs, where the woodland pasture species are most abundant. The scenic beauty of the Palisades is a constant draw for the development of housing estates. A long-term effort by the Nature Conservancy, the Kentucky State Nature Preserves Commission, and other organizations to protect these precious lands has slowed development. Today there is an extensive network of nature preserves and other protected land that, although not sufficient, is essential to the continued existence of the forests that provide habitat for the woodland pasture species.

Away from the major rivers, creeks meander through the Bluegrass. Some, like the Elkhorn and Town Branch, figure prominently in the history of the Bluegrass. The Elkhorn was an important route to approach Lexington from the Kentucky River at what was then known as Leestown. Lexington developed around Town Branch, which is now mostly buried beneath pavement, emerging only as it leaves the downtown area.

One searches largely in vain for woodland pasture species along these

Figure 2.12. The upper banks of creeks in the Inner Bluegrass were once lined with giant cane as tall as thirty feet, but today the cane has been replaced with a mixture of small bottomland trees and invasive species such as bush honeysuckle and wintercreeper.

Figure 2.13. Reforest the Bluegrass is a volunteer program to replant native trees along Inner Bluegrass creeks to improve water quality. The black squares protect the young trees from grass competition. Volunteers plant woodland pasture species such as bur oak and Shumard oak on the upper slopes, and bottomland species on the creek banks.

Figure 2.14. A young volunteer planting bur oaks on the upper slope of West Hickman Creek for Reforest the Bluegrass.

upland creeks. The creek banks have been highly disturbed by centuries of grazing, industry, and development. Most of the creeks are lined with nonnative species such as fescue (*Festuca* spp.), bush honeysuckle (*Lonicera* spp.), and wintercreeper (*Euonymus fortunei*), which prevent native trees from becoming established. In recent years, there have been major efforts to reforest the banks of the creeks, which have met with some success. Reforest the Bluegrass is one of the largest volunteer tree-planting programs in the United States. Its purpose is to restore forests to riparian (stream-bank) areas to improve water quality. Wetland tree species are planted near the creeks, whereas the slopes above the creeks are well suited to woodland pasture species such as bur oak and Shumard oak.

How Much Is Left?

We know from the work that went into preparing this book that there are thousands of ancient trees in the Bluegrass, many of them dating to times before any Europeans arrived here. Others appear to be a bit younger, as if natural regeneration of woodland pasture species continued for some time after settlement before stopping by the late 1800s. We don't know why natural regeneration stopped, although it is most likely due to an increase in grazing intensity and the introduction of mechanized farm equipment.

Today there is almost no natural regeneration of the woodland pasture species except along the rivers. While many farm owners plant trees, they are usually ornamental selections that do not represent the original woodland pasture species. The majority of ornamentals being planted are short-lived species that could never replace the function of woodland pastures, even if they provide some temporary shade.

We do not know how many acres of original woodland pastures remain. Some of the woodland pastures cover hundreds of acres, crossing property boundaries. The land is almost entirely private, and access to it for inventory purposes has so far been limited. Aerial photography does not help, because it has not been possible to distinguish native woodland pasture species from planted ornamentals.

Although we may not know how many acres of original woodland pasture remain, there is no doubt that woodland pastures are in steep decline, and that we will lose them completely if we do not act to conserve them.

3

The Llama Tree

Venerable Trees in the Nashville Basin

The llama tree shades a small pasture in Tennessee near College Grove. Two llamas sit in the shade of the tree, roll around in the dust nearby, and occasionally snack on the tree. It is a huge bur oak, one of many ancient trees scattered throughout the farmland of the Inner Nashville Basin. Nearby, a huge, dead bur oak stands at the edge of a pasture occupied by longhorn cattle. At nearly every turn of a country road, solitary old trees stand in the pastures.

It is possible to find woodland pastures of ancient trees in the Nashville Basin, but they are much less common than in the Bluegrass. Yet the abundance of solitary ancient trees throughout the Nashville Basin tells us that there were woodland pastures here hundreds of years ago.

The Nashville Basin is divided into an Inner Basin and an Outer Basin. The Inner Basin consists of level to gently rolling land with shallow soils that are low in phosphorus. Livestock farms are small and devoted largely to cattle. Cedar glades have formed on the thin soils and exposed limestone rock. Cedar glades are open areas containing a rich diversity of uncommon plant species, including rare and endangered species. The glades are dominated by herbaceous plants, but trees are found in cracks in the exposed limestone, and forests form around the edges. Eastern redcedar is the most common tree of the glades, but woodland pasture species are represented as

Figure 3.1. The llama tree is an old bur oak shading a small pasture on a farm near Shelbyville, Tennessee, providing shade and snacks for the resident llamas.

Figure 3.2. Near the llama tree, a dead bur oak provides a perch for a turkey vulture overlooking long-horned cattle.

Figure 3.3. A woodland pasture in the Inner Nashville Basin includes the same species that grow in the Inner Bluegrass, but there are more shagbark hickory trees here.

well. Blue ash and Shumard oak are common around the glade margins. The cedar glade habitat is found in western Kentucky, but not in the Bluegrass, where the soils are deeper and more fertile.

The Outer Basin is complex, comprising level pasture land and forested hills. The level land, especially south of Nashville, is fertile, its phosphorus-rich soils like those of the Inner Bluegrass. The hills are forested, although much of the land is now urban.

The Nashville Basin and the Bluegrass have the same geological origin, the uplift of an arch of limestone followed by erosion into level basins. The Inner Nashville Basin is almost entirely Ordovician limestone, whereas the Outer Nashville Basin is a mixture of different kinds of limestone, mostly in the form of coarse-grained calcarenite that is rich in phosphorus. The hills of the Outer Nashville Basin are a complex mixture of shale, limestone, and chert. This mixture is important for our understanding of the distribution of woodland pasture tree species in the Outer Basin.

Land use in the Nashville Basin is quite different from that in the Bluegrass. Nashville has grown into a metropolitan area of 1.7 million people in sixteen counties, including nearly all the counties of the Inner and Outer Basins. As Nashville has grown and blended with surrounding cities

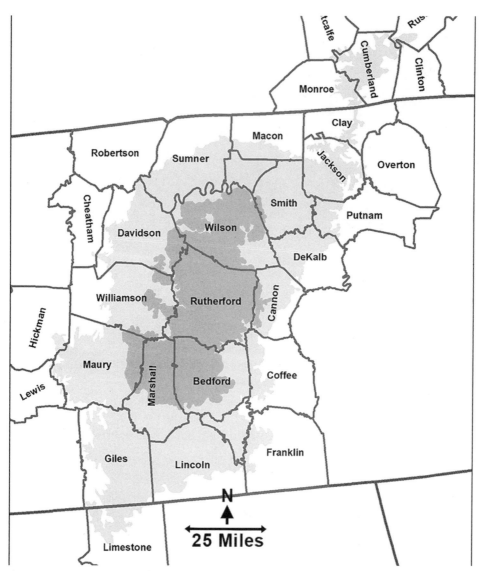

Figure 3.4. Ecoregions of the Nashville Basin. The Inner Nashville Basin (dark gray) is similar to the Inner Bluegrass, with rolling hills, farms, and a few woodland pastures of venerable trees. Soils here are thinner and less fertile than in the Inner Bluegrass. The Outer Nashville Basin (light gray) is rolling and hilly; woodland pasture species are restricted to the lower, more level land, and the hills are generally forested.

like Murfreesboro and Franklin, development has engulfed much of the farmland.

In spite of its growth as a major metropolitan area, Nashville and its surrounding counties are remarkably lush and green. This is evident from the urban forest canopy cover of Nashville. The health of an urban forest is often measured by its canopy cover—the amount of land surface covered by

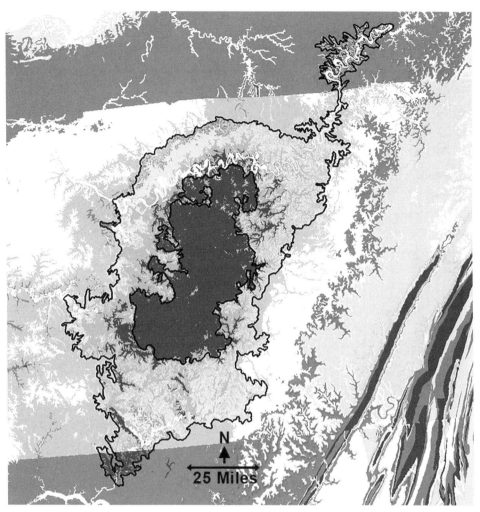

Figure 3.5. Karst in the Nashville Basin. Limestone is the dominant rock only in the Inner Nashville Basin. The Outer Nashville Basin is more dissected with karst mostly on the lower elevations and level ground.

tree canopies. The national average for tree canopy cover is 27 percent, but the goal for healthy cities in the eastern United States is to achieve 40 percent cover. Nashville has 47 percent canopy cover—nearly half of the city is shaded by trees. By comparison, Lexington has a canopy cover of 25 percent in the urban service area.

Much of the lush, green cover of Nashville is found in its vast parks. The largest parks occupy the hills in the Outer Basin. The Warner Parks are a single continuous forest area of more than 2,600 acres. The nearby Radnor Lake State Natural Area encompasses 1,200 acres.

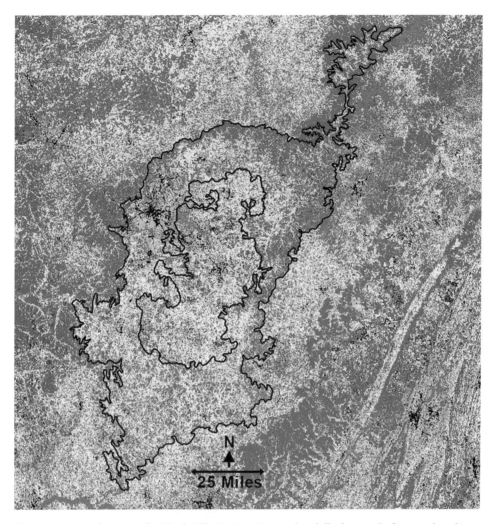

Figure 3.6. Land use in the Nashville Basin. Pastureland (light gray), forests (medium gray), and urban areas (dark gray) create a more complex mosaic than is seen in the Bluegrass. Pastures are generally smaller than in the Inner Bluegrass.

My family often spends a week or two in Nashville in the summer. While our kids attend camps and courses at Vanderbilt University, my wife and I have some free time. Neither of us is especially drawn to city life, so we spend our vacation days exploring the green spaces in and around Nashville. I have also spent a number of days exploring the rural areas of the Nashville Basin looking for woodland pastures and ancient trees.

The llama tree is typical of most of the venerable trees of the Nashville Basin. Large woodland pastures are uncommon, but there are many individual ancient trees like the llama tree. The largest trees are almost always

Figure 3.7. Bur oaks are common as solitary old trees in developed areas throughout the Nashville Basin. Chinkapin oaks are also fairly common as individual trees.

the same species we see in the Bluegrass. Among the largest trees, bur oak is far more abundant than the others. Within a mile or so of the llama tree are at least ten more bur oaks, but few of the other species. Big chinkapin oaks can be found scattered throughout the Nashville Basin, as are large Shumard oaks, but the largest trees are mostly bur oaks. Small blue ash trees are abundant around the cedar glades, but big, old blue ash are not as common here as they are in the Bluegrass.

With sufficient exploration, a pattern begins to emerge, though I do not profess to know the Nashville Basin as well as the Bluegrass. In the Inner Nashville Basin, huge, old trees are dotted around the landscape, occasionally in small groves, much less commonly in large woodland pastures. And those woodland pastures are different from the ones in the Bluegrass. The soils are thinner, and there are several species, such as white oak and shagbark hickory, that we do not find in Bluegrass woodland pastures.

The Outer Nashville Basin has richer soil and more productive farms, but also more development. The southern suburbs of Nashville include some of the wealthiest neighborhoods in the country, characterized by huge houses on large estates. It is here that we find the greatest abundance of

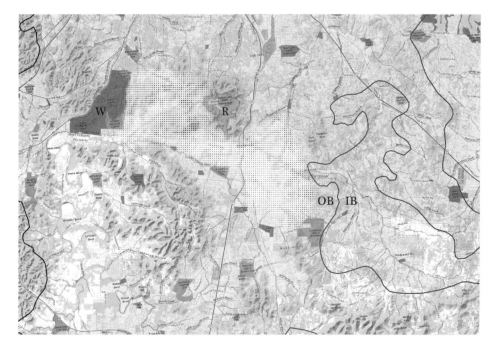

Figure 3.8. Southern Nashville. Key: W—Warner Parks, R—Radnor Lake, OB—Outer Basin, IB—Inner Basin. The stippled area is where we find the largest number of ancient bur and chinkapin oaks today, scattered throughout farms and suburban developments.

Figure 3.9. A bur oak southeast of Nashville. This is the southernmost large bur oak I have found in the Nashville Basin.

venerable tree species, dominated by bur oak. Many of these properties are off-limits to visitors, but there are enough small roads and broad vistas to get a sense of the abundance of woodland pasture tree species.

The Warner Parks, consisting of Percy Warner Park and Edwin Warner Park, are quite interesting. They consist of second-growth forest on hills to the south and west of the city. The dominant rock is shale, with a hard cap of chert, but limestone outcrops are found throughout the parks. On these limestone outcrops we can find all the trees of the woodland pasture—bur oak, chinkapin oak, Shumard oak, kingnut, and blue ash. Farther east, Radnor Lake State Nature Area is lacking only bur oak.

In between these parks, on the level ground in Brentwood and east toward the Inner Basin, is where we find the greatest number of large, old trees that are characteristic of woodland pastures. I suspect that, before settlement of this area, there were extensive woodland pastures on the level ground from the Warner Parks sweeping south and east. Brentwood was settled at about the same time as Lexington. As in Lexington, many of the farms were very large estates, but after the Civil War, as the southern economy declined, the farms were broken up into smaller and smaller farms. Today only the occasional giant bur oak or chinkapin oak reminds us of what might have been here.

Belle Meade is a small town at the base of the Warner Parks. It was once a 5,400-acre plantation and renowned Thoroughbred farm that grew from a small horse farm on the Natchez Trace. Like many plantations, Belle Meade declined after the Civil War, and most of the land was sold for development, leaving today's thirty-acre Belle Meade Plantation surrounded by fine houses. I have found a few old bur and chinkapin oaks, along with quite a few large, old white oaks in Belle Meade. I suspect that Belle Meade Plantation bore a close resemblance to some of the horse farms of the Bluegrass, its intact woodland pastures lasting up until the Civil War, but I have not yet found historical descriptions of the vegetation.

Whatever the condition of woodland pastures in this area at the time of settlement, the end of the Civil War was devastating to the agricultural economy of the region. Plantations were divided into smaller farms, and it is likely that most of the land was cleared at that time for more intensive agriculture. I am speculating here, but it seems likely that the Civil War played a major role in the loss of woodland pastures. With the growth of the Nashville area, development would have eliminated what remained of the woodland pastures, leaving scattered trees.

Figure 3.10. The Cumberland River in the Nashville Basin. The Cumberland does not cut deeply into the limestone, and it lacks the cliff-and-bluff habitat of the Kentucky River. Few woodland pasture species are found along the Cumberland owing to disturbance.

The Nashville Basin is the southern limit of the range of blue ash and kingnut, and the southeastern limit for bur oak. A magnificent bur oak southeast of Nashville is the most southerly large bur oak I have been able to find. Trees at the limit of their range are important, especially in today's era of rapid climate change. Trees at the southern end of their range are likely to die as ranges shift northward. On the other hand, trees like this one, which has obviously stood through heat and drought for hundreds of years, may be important to the survival of bur oaks in the future. It is probable that these southerly individuals are genetically more tolerant of drought and heat than their more northerly cousins. Bur oaks in particular are largely northern trees. If this tree is genetically more tolerant of drought and heat, seeds collected from it may make an important contribution to the future of the species.

4

Venerable Tree Species

There are many species of trees native to the Bluegrass and Nashville Basin. The majority are found in forested areas along the rivers and steep slopes. Fewer species are found in our woodland pastures, and they share some common characteristics. With a few exceptions, they are all very drought tolerant, tend to be deep rooted, and either are found only on soils derived from limestone or prefer such soils.

We cannot make an authoritative list of the trees that were in our woodland pastures in 1779. There were no botanists here when the first settlers arrived. The settlers were pragmatic people who could distinguish among many species on the basis of their utility. So, for example, they knew the difference between white oaks and red oaks, but within the red oak group they would not have distinguished among species because that knowledge was not useful to them. The names of trees could also be confusing. Hackberry was variously referred to as hackberry, sugarberry, and hooptree, but these terms could also be applied to other trees. Sugartree could refer to sugar or black maple, from which sap was obtained for boiling into sugar, or to hackberry or sugarberry, from whose fruits sugar could be extracted.

Nevertheless, from multiple accounts by early settlers, from botanists who came later, and from our existing woodland pastures, we have a fairly good idea of the upland trees that were here in 1779 and earlier.

What follows is an introduction to the five species that form our current woodland pastures and that are long-lived enough to have been here in 1779. There are other species that were here in 1779, and whose progeny are here today, and we will briefly describe them. This chapter is not intended as a guide to identification, for which the reader is referred to the several excellent field guides for the region.[1]

Because our trees are large and old, we list what we know about the oldest and largest trees of each species. This information should not be regarded as definitive. Determining the age of trees is not a trivial task, and we are constantly finding trees much older than we thought possible for a given species.

Each species description is accompanied by a map that shows the species' range before European settlement. The maps also show importance value, which is a measure of the abundance and size of trees of each species compared with other trees in today's forests. Darker squares indicate greater abundance of the species. The range maps provide a historical view of where each species might have been found around the time of European settlement, and the importance value provides a snapshot of the relative abundance of each species in today's forests. The range maps were created by Elbert Little of the U.S. Forest Service. The importance value is calculated from the U.S. Forest Service Forest Inventory and Analysis Program.[2]

Bur Oak

- Common names: Bur oak, burr oak, mossycup oak, blue oak, scrub oak, mossy overcup oak[3]
- Botanical name: *Quercus macrocarpa* Michx., Fagaceae, the beech family[4]
- Largest bur oak: 94 inches diameter, 99 feet tall, Indiana. Largest Bluegrass bur oak: 92 inches diameter, 102 feet tall, Woodford County
- Oldest verified bur oak: 401 years, Custer State Park, South Dakota; bur oaks estimated at 500 years have been found in the Bluegrass

Everything about a bur oak is massive. The whole tree appears heavy: a stout stem, thick, wandering branches, and very large leaves. Even the fruit is huge, the largest of any oak tree. The Latin name *Quercus macrocarpa* means large-fruited oak. The terms *bur, burr,* and *mossycup* refer to the fringe around the acorn cup.

Bur oaks are found in closed forests along streams, but in our region are found mostly as open-grown trees with huge, spreading crowns. Bur oak is intermediate in shade tolerance, meaning that it can grow in partial shade, especially when young. In spite of its shade tolerance, bur oak is usually a widely spaced, open-grown tree.

Figure 4.1. The range and importance value of bur oak. Range is indicated by the black outlines. Importance value is a relative measure of species abundance and tree size in today's forests. Importance is indicated by gray squares; darker squares indicate higher importance value. Trees may still be present in low numbers in areas with no squares.

Bur oak has a large range, from Maine and New Brunswick to Saskatchewan and Texas. Today it reaches its greatest abundance in Minnesota, Iowa, Wisconsin, and Kansas.

In the upper Midwest, where it is most abundant, bur oak is the characteristic tree of the savanna that once formed a broad divide between the tallgrass prairies to the west and the deciduous forests to the east. There is little doubt that fire set by Indians played a critical role in creating and preserving the oak savannas of the Midwest. Today oak savannas occupy a tiny remnant of a once-vast ecosystem. Bur oak, with its thick bark, corky branches, and deep rooting, is the most fire-tolerant oak species, and this has been important in its success in savannas.

In the Bluegrass and Nashville Basin, the fire tolerance of bur oak is not as important as its tremendous drought tolerance. Bur oaks often

Figure 4.2. Leaves of bur oak.

become established during droughts, which reduce competition for light from grasses and cane. These drought-initiated bur oaks can be extremely deep-rooted. I have seen bur oaks with no roots within the top six feet of soil. These trees are likely to have most of their roots deep in cracks in the fractured limestone.

Bur oak is a forest tree along river bottoms and low ridges in the western part of its range. In the Bluegrass and Nashville Basin, bur oak is occasionally found in woods along creeks, but is more likely to be encountered in woodland pastures.

Bur oak is certainly one of the most recognizable and characteristic trees of woodland pastures in the Bluegrass and Nashville Basin. It is so prominent in our landscape that almost everyone recognizes it, although it is not as abundant as blue ash or chinkapin oak.

As Josiah Collins recounted, bur oak was quite useful as a timber tree to early settlers, and a lot of them were cut down for building material. It was also valued for its shade, however, and many were left standing to survive until today. There is some natural regeneration of bur oaks, and younger trees are often found along fencerows.

Bur oak can be purchased from some tree nurseries. The acorns for many nursery trees come from Lexington Cemetery and the University of

Figure 4.3. A bur oak in a woodland pasture.

Kentucky campus. Bur oak should be planted only in large areas with plenty of room and soil that is not compacted. It should never be planted as a street tree, because the narrow grass strips between the sidewalk and the street do not provide adequate rooting volume for such a large tree.

Blue Ash

- Common names: Blue ash, square-twig ash
- Botanical name: *Fraxinus quadrangulata* Michx., Oleaceae, the olive family
- Largest blue ash: 66 inches diameter, 82 feet tall, 72.5 inches crown spread, Jefferson County, Kentucky
- Oldest verified blue ash: 249 years, Harrison County, Kentucky; oldest estimated blue ash: 405 years, Inner Bluegrass

Blue ash gets its name from a blue dye that is made by soaking the inner bark or twigs in water. Native Americans and pioneers used blue ash to dye fabric. Its botanical name, *Fraxinus quadrangulata*, is Latin for "four-angled ash," a reference to its squarish twigs. Blue ash is a member of the olive family.

Blue ash is the most abundant tree of our woodland pastures, and in

Figure 4.4. The range of blue ash. See Figure 4.1 for an explanation of the map.

Figure 4.5. A blue ash in a woodland pasture.

many places it is the only species remaining. Blue ash is an open-grown tree of limestone uplands, in pure stands or mixed with other woodland pasture species. It is not a very neat tree, its branches irregular and coarse. Old trees almost always have signs of having lived a tough life, from dead branches to tops blown out by lightning, which leave a characteristic stag-head

Figure 4.6. Blue ash has square twigs, broadly winged fruit, and pale foliage.

appearance. Blue ash has never achieved the ornamental prowess of white ash and is not generally available from nurseries.

Blue ash is a long-lived, slow-growing tree. It is moderately tolerant of shade, and seedlings may persist for a time under a moderate forest canopy. Seed production is high, but germination rates are very low, typically less than 7 percent, because of complex dormancy and weevils. Reproduction appears to be difficult for this species in its current habitats. Seedlings and saplings can be found in bluffs of the Kentucky River, along creeks, and in a few nature preserves, but they are very rare on Bluegrass uplands. Blue ash reaches its greatest abundance in the Inner Nashville Basin, where it is a very common large tree in woodland pastures and around the margins of cedar glades.

Blue ash is very different from the other ash species in appearance and habitat. It is also quite distinct genetically, having no close relatives in eastern North America, and does not hybridize with other ash species.

The emerald ash borer (*Agrilus planipennis*), an Asian beetle, is currently devastating populations of ash trees throughout eastern North America. Introduced in or near Detroit in the 1990s, the insect is currently killing ash trees in the Bluegrass and becoming established in the Nashville Basin.

White ash and green ash have little or no resistance, and the majority of these trees will be killed in the next few years. Blue ash is somewhat tolerant of emerald ash borer. Sara Tanis and Deborah McCullough, entomologists at Michigan State University, surveyed ash trees in southern Michigan after the emerald ash borer had done its dirty work. They reported that the majority of blue ash trees survived attack, but all white ash were killed. It remains to be seen whether this resistance is sufficient to allow blue ash trees to survive in the Bluegrass and Nashville Basin. Landowners are still advised to treat blue ash trees to kill the emerald ash borer until we have more certainty about their fate.[5]

In addition to its use as dye, blue ash was a popular timber tree for a time. Many old homes in the Bluegrass and Nashville Basin have blue ash flooring, molding, and furniture.

The blue ash trees in our woodland pastures today appear to be a mixture of very old trees and somewhat younger trees. It seems that natural reproduction of blue ash continued for some time after the development of farmlands, perhaps for as long as one hundred years. The tree, however, is not reproducing to any extent in woodland pastures today.

There are no authentic cultivars of blue ash—it is difficult to grow from cuttings, and "blue ash" cultivars on the horticulture market generally prove to be other species. Commercial nurseries avoid all ash trees at present because of emerald ash borer.[6]

Shumard Oak

- Common names: Shumard oak, Shumard's oak, spotted oak, Schneck oak, Shumard red oak, southern red oak, swamp red oak
- Botanical name: *Quercus shumardii* Buckley, Fagaceae, the beech family. Named for Benjamin Franklin Shumard (1820–1869), a physician and geologist
- Largest Shumard oak: 91 inches diameter, 136 feet tall, Clay City, Kentucky; largest Shumard oak in the Bluegrass: 92 inches diameter, 134 feet tall, Lexington
- No verified oldest Shumard oak; oldest Bluegrass Shumard oak estimated at 480 years

Shumard oak is the most confusing and vexing tree in the Bluegrass. Like other red oaks, Shumard oak can be very difficult to identify in the field. Early settlers found red oaks here, but they did not distinguish among the

Figure 4.7. The range of Shumard oak. See Figure 4.1 for an explanation of the map.

many species. Other red oaks that can be found in the Bluegrass and Nashville Basin are black oak, northern red oak, pin oak, scarlet oak, shingle oak, and southern red oak.

Shumard oak is found throughout the Bluegrass and Nashville Basin, largely in the forested slopes and river valleys. It is one of the most common trees on slopes above the Kentucky, Licking, Salt, Ohio, and Cumberland rivers. It can be found in the woodland pastures but is less common than the other species. This may indicate more its usefulness as a timber tree than its original abundance: many Shumard oaks were probably felled for timber. Farther south, Shumard oak is a forest tree found mostly on river terraces where there is rich, moist, but well-drained soil. Although it is often found near rivers, it is not found in the very wet bottomlands.

A careful look at red oaks in the Bluegrass shows that there are many individual trees that have characteristics intermediate between two or more species. I have long suspected that Shumard oak in the Bluegrass is either

Figure 4.8. A large Shumard oak in a park.

Figure 4.9. Leaves and acorns of Shumard oak.

a different species from Shumard oak farther south, or that the Bluegrass is the center of a hybrid zone where many trees bear genes from parents of different species. Shumard oak can hybridize with all the red oaks listed above. I quite commonly encounter trees that are intermediate between pin oak and Shumard oak and between northern red oak and Shumard oak. Whatever the true nature of the parents of these trees, it is often not possible to assign a particular tree with certainty to one species or another. Genetic analysis of oaks shows that, though hybrids among species are common, each species maintains its identity.

Shumard oak does not grow well in shade, but like the other Bluegrass oaks, its seedlings may persist in the forest understory until an opening in the canopy allows them to grow more quickly. On good sites, Shumard oak is faster growing than chinkapin oak or bur oak. Like most oaks, Shumard oak is a mast-fruiting species that bears abundant crops of acorns at irregular intervals. Shumard oak seedlings and saplings are common around the edges of forests—for example, on the bluffs above the Kentucky River and on the slopes of the Cumberland River—but they do not reproduce in woodland pastures except along fencerows. The longevity of Shumard oak is not known, but many Bluegrass Shumard oaks are probably more than 200 years old, and one tree was estimated to be 480 years old.

Shumard oak is readily available in nurseries and is a common street and park tree in our area. These, however, are ornamental cultivars from farther south, especially from Texas, and it is not clear that they are the same species or that they are as well adapted to our conditions as native Shumard oak trees. In the interest of conserving woodland pastures, it will probably be wise to grow seedlings from local seed sources.

Chinkapin Oak

- Common names: chinkapin oak, chinquapin oak, yellow oak, rock oak, yellow chestnut oak
- Botanical name: *Quercus muehlenbergii* Engelmann, family Fagaceae, the beech family. Named for Gotthilf Heinrich Ernst Muhlenberg (1753–1815), a pastor and botanist in Pennsylvania.[7]
- Largest chinkapin oak: 99 inches diameter, 76 feet tall, Griffith Woods, Harrison County, Kentucky
- Oldest verified chinkapin oak: 429 years, Guadalupe Mountains National Park, Texas; oldest verified Bluegrass chinkapin oak: 403 years, Fayette County, Kentucky

Figure 4.10. The range of chinkapin oak. See Figure 4.1 for an explanation of the map.

Chinkapin oak gets its common name from a Virginia Algonquian word, *chechinquamin*. The exact meaning is unknown (Virginia Algonquian is an extinct language), but it may have been used for the fruit of several species of chestnut. There are several other small trees and shrubs called chinkapin or chinquapin that have fruits similar to the chestnut's. It is likely that chinkapin oak was given its name because its acorns are sweet like chestnut seeds and were desirable as a food source to the Algonquin and other Native people.

Of the three oak species found in our woodland pastures, chinkapin oak is the most common. It is also an abundant component of the denser forests along the Kentucky, Licking, Salt, Ohio, and Cumberland rivers in the Bluegrass and Nashville Basin. Chinkapin oak is almost always associated with limestone and is rarely found on soils derived from other rock types. It is quite drought tolerant and able to withstand the rigors of life on our droughty karst landscape.

Chinkapin oak grows slowly and lives for a very long time. It is

Figure 4.11. A large chinkapin oak in a woodland pasture.

intolerant of shade, though it can persist in a shady understory when young. Chinkapin oak produces abundant acorns at irregular intervals, but reproduction is a rare event in many areas. In the Bluegrass, we see very little natural regeneration of this species in woodland pastures, but it is fairly common on bluffs and the upper slopes above rivers.

Figure 4.12. Leaves of chinkapin oak.

Open-grown trees are massive, their trunks very thick. Their crowns are finer than those of bur oak, and the tree generally has a more elegant appearance. In forests, chinkapin oaks are tall, dominant canopy trees.

Hybrids with other white oak species are fairly common, including hybrids with white oak (*Quercus alba*), swamp white oak (*Quercus bicolor*), and bur oak (*Quercus macrocarpa*). Deam oak (*Quercus muehlenbergii* × *alba*) is rare, but I have seen several in the Bluegrass that appear to be very old.

Chinkapin oak seedlings are occasionally available from state nurseries, but only rarely from commercial nurseries. Its very slow growth rate limits the commercial market for chinkapin oak. It is a beautiful ornamental if it has plenty of space and time, but it is not suitable as a street tree.

Kingnut

- Common names: kingnut, shellbark hickory, big shagbark hickory, bigleaf shagbark hickory, bottom shellbark, thick shellbark, western shellbark

- Botanical name: *Carya laciniosa* (Michx. f.) G. Don, Juglandaceae, the walnut and hickory family
- Largest kingnut: 56 inches diameter, 139 feet tall, Greenup County, Kentucky
- There are no verified age measurements available, but kingnut is known as a long-lived tree, and there are many trees in the Bluegrass that are probably over three hundred years of age

Kingnut is the name given to this tree by early settlers. The huge nut, the largest of any hickory, was a significant food source for Native people and settlers. Later, botanists began using other names, such as shellbark hickory. I will stick with kingnut because it is historically significant, and because the name shellbark is easily confused with shagbark.

Hickories are among the most common trees of the Bluegrass and Nashville Basin, including pignut, red pignut, shagbark hickory, bitternut, and kingnut. Of these, kingnut is the most common in woodland pastures,

Figure 4.13. The range of kingnut. See Figure 4.1 for an explanation of the map.

Figure 4.14. The characteristic shaggy bark of kingnut.

Figure 4.15. Kingnut leaves and fruit.

while the others tend to occur on forested slopes. Everything about this tree is stout—from its massive trunk, thick twigs, and large leaves to the huge fruit and nut. A large, open-grown kingnut in a woodland pasture is very impressive.

On limestone soils, kingnut is deeply rooted and drought tolerant. Outside our limestone habitats, kingnuts can be found along forested streams and are replaced by shagbark hickory on drier upland sites. It is also found in bottomlands and swamps in the Midwest.

Kingnut is a slow-growing, long-lived tree. Like other hickories, it is mast-fruiting, producing abundant crops at irregular intervals. The huge nuts are dispersed by squirrels and gravity. Squirrels often abandon kingnuts after chewing on them for a while—the husks and hulls may be too thick for all but the most dedicated squirrel. The nuts germinate in spring, and a deep taproot develops quickly in good soil. Shoot growth is slow, and even on good sites the tree may reach only two feet in height in five years.

Kingnut is quite tolerant of shade and can reproduce in fairly dense forests. Even with adequate sunlight, it grows more slowly than associated species. Kingnut is very susceptible to fire damage, and it is not found in oak savannas and other forests where fire is an important component.

Kingnuts were important food sources for Shawnee and other people and for early settlers and their livestock. Though rarely harvested today, the nuts are large and delicious.

Kingnut is reported to hybridize with pecan (*Carya illinoinensis*) and shagbark hickory (*Carya ovata*). In the Bluegrass, kingnut and shagbark hickory are easy to distinguish, and I have never seen a hybrid tree.

Kingnut is rarely available from nurseries, as it is slow growing and difficult to transplant. Because it is deeply taprooted, kingnut may be better suited to direct seeding—planting the seeds where a tree is desired—rather than transplantation from a nursery.

OTHER TREES IN THE BLUEGRASS AND NASHVILLE BASIN

The five species described above, which we refer to collectively as the venerable trees, are of course not the only trees in the region, and not even the most common. They form the core species of the woodland pastures and are the ones we are sure can reach great age. Many of the individual trees of these species were here in 1779 and before.

There is another group of trees that were present in woodland pastures in 1779 and are here today, but today's trees are the descendants of the original trees, not the same individuals. These are trees that either are naturally short-lived or were damaged or harvested by settlers. They are listed in approximate order of abundance in today's woodland pastures.[8]

Hackberry (*Celtis occidentalis*) is the most abundant tree in the Inner Bluegrass and possibly in the Nashville Basin. It is a very fast-growing tree and does not reach great age. Large hackberries were prized for furniture because the wood is soft and easy to work, so the big trees were harvested soon after settlement. Most hackberries in our area are less than one hundred years old. Hackberries still come up everywhere except in mowed pastures. Hackberry is abundant on fencerows, on forest edges, and in cities.

Black walnut (*Juglans nigra*), a very common and important tree often found in woodland pastures, does not live to a great age. Black walnut is highly valued as a timber tree, and most that were here at the time of settlement were harvested. Today black walnut is found in most of our woodland pastures and urban areas. It reproduces well along fencerows and in unmowed pastures. Black walnut is also common in urban areas, parks, yards, and housing developments. It is threatened by a very serious insect and disease complex called thousand cankers disease. Although not at press time reported in the Bluegrass or Nashville Basin, the disease is in surrounding regions and is likely to be here within a few years if it is not already. The future of black walnut in our region is uncertain.[9]

Black locust (*Robinia pseudoacacia*) was common in the Bluegrass at the time of European settlement, and it was often reported to be a very large, straight tree. Black locust is a very fast-growing, short-lived tree; although common in woodland pastures today, it is unlikely that there are any old trees. Black locust can sprout from its roots, creating a stand of genetically identical trees, known as a genet, that can occupy a fairly large area. The root system of the stand can be very old even though individual stems are not. Black locust was valued for making fence posts. Although we make little use of the tree today, it is the most widely planted North American tree in Europe and southern Africa, where it is often an invasive nuisance tree.

Hickories (*Carya* spp.) other than kingnut are often found in woodland pastures and are abundant in woodlands. The most common Bluegrass hickories are bitternut (*Carya cordiformis*), pignut (*Carya glabra*), and red pignut (*Carya ovalis*). Less common are shagbark hickory (*Carya ovata*) and mockernut (*Carya tomentosa*). In the Nashville Basin, shagbark hickory is more abundant than kingnut in some woodland pastures.

Honeylocust (*Gleditsia triacanthos*), like its close relative black locust, was present in significant numbers. Settlers prized the fruit of honeylocust and often commented on the presence of the tree. At a time when sweeteners like sugar and honey were very expensive, the gum inside a honeylocust pod was a source of both sugar and a thickening agent. The gum was used for making a variety of deserts and was the main sweetener in persimmon beer, a popular beverage.

Honeylocust is common along streams and rivers and on bluffs above the rivers, but in the Bluegrass it is found on uplands, especially abandoned pastures and dry woods. Honeylocust can reach great size, though we don't often find big ones in our area, and it is unlikely that there are any presettlement honeylocust trees present. A thornless variety of honeylocust is very popular as a shade tree, and is abundant in our cities. Seedlings from the thornless trees revert to the wild, thorny variety.

Ohio buckeye (*Aesculus glabra*) is very common in limestone habitats and can reach substantial size. I find Ohio buckeye in about a third of the larger woodland pastures, and it is probably present on every farm, along hedgerows and creeks. The largest Ohio buckeye known is in a woodland pasture remnant in Lexington, but we do not know its age. Yellow buckeye

(*Aesculus flava*) is found on slopes near rivers in our area, as are hybrids between the two species.

American elm (*Ulmus americana*) is still common in the Bluegrass in spite of the depredations of Dutch elm disease. It is an uncommon component of the woodland pastures and is found more often along the major rivers. Large trees are still found in the Bluegrass and Nashville Basin.

Slippery elm (*Ulmus rubra*) was probably more abundant than American elm, and most early settlers seem to have recognized the difference because of the important medicinal properties of slippery elm—the inner bark is soothing to irritated throats. Early accounts related that once cattle were introduced, they quickly chewed the bark from the slippery elms, something they did not do to other trees. It is likely that cattle killed any slippery elm trees in woodland pastures. Today slippery elm is found mostly along hedgerows and creek banks. Rock elm (*Ulmus thomasii*), a species almost always associated with limestone, is present but not common in the Bluegrass, mostly along creeks or the margins of woodland pastures. In the Nashville Basin, rock elm is a common component of dry, rocky limestone habitat.

Kentucky coffeetree (*Gymnocladus dioicus*) is one of the most recognizable trees in our area and is found in woodland pastures and around the edges of forests throughout the Bluegrass and Nashville Basin. As individuals, coffeetrees probably do not reach great age, and there are unlikely to be any presettlement trees still living. Coffeetree can root-sprout and produce clones, often in fairly precise circular "fairy rings." It is possible that there are old clones of coffeetree, but individual stems are not very old. Coffeetree is a beautiful ornamental for large sites, especially in winter, when its extraordinarily stout branches are bare. Coffeetree got its name from the resemblance of the seeds to coffee beans, and some narratives say that early settlers used it as a coffee substitute. I think this unlikely and can attest from personal experience that it makes an awful coffee substitute.

White ash (*Fraxinus americana*) was common at the time of settlement, and it remains one of our most abundant trees, often forming pure stands on abandoned farmland. It is a fast-growing and fairly short-lived tree, probably not living more than two hundred years. Although there are large white ash trees in the Inner Bluegrass, it is unlikely that any of them are presettlement trees. Today white ash trees are being killed by emerald ash borer, and most experts believe that nearly all the white ash in Kentucky and Tennessee will be gone in a few years.

Yellow-poplar (*Liriodendron tulipifera*) is also called tuliptree or tulip-poplar, or just poplar. The state tree of Kentucky and Tennessee, yellow-poplar is an extremely important tree in Appalachian forests. It was often reported in early accounts of the Bluegrass because it is such a useful tree. It was probably not common, however, and was restricted to forests along slopes and river corridors, not as a component of the woodland pastures. Yellow-poplar wood is soft and workable and made fine molding, flooring, and furniture, so even if the trees were present in woodland pastures, they would probably have been harvested. Yellow-poplar is both very fast growing and long-lived, reaching at least five hundred years of age in the Appalachians. There are a few very large yellow-poplars in and around old estate homes and cemeteries, but they were probably planted after settlement.

Sugar maple (*Acer saccharum*) and black maple (*Acer nigrum*) are abundant in woodland pastures but are rarely old. I have not seen a single sugar maple or black maple in woodland pastures that has the characteristics of a very old tree. Yet both species can live to great age. The lack of old trees probably has to do with their use by the first farmers. Early narratives consistently reported on the abundance of these "sugar trees" and their importance to settlers as a source of sugar, although some of these accounts were referring to sugarberry (*Celtis laevigata*). The method used to obtain sap from sugar maples was not the gentle tapping used in New England, however; rather, axes cut deep gashes in the trees' bark. The trees did not survive the insult for very long, and today there are no more than a few sugar maples that might be over two hundred years old. Early accounts do not distinguish between sugar maple and black maple, but both species are found today in many woodland pastures.

Black cherry (*Prunus serotina*) is common in woodland pastures and hedgerows. Although the trees do not have the fine form of northern black cherry, they were valuable for their wood and harvested quickly. Black cherry is not a long-lived tree, and it is unlikely that any presettlement black cherries are present today.

Eastern redcedar (*Juniperus virginiana*) can undoubtedly reach an astonishing age, like other members of its family. The oldest verified redcedar in West Virginia is 940 years old. Eastern redcedar is such a valuable tree for closets, chests, and animal bedding that it was all harvested very soon after settlement. Today the species is making a comeback with the abandonment of farmland, particularly in the Hills of the Bluegrass and the Inner Nashville Basin. It is possible that a few very old redcedars are clinging to the Kentucky River Palisades. Eastern redcedars are occasionally found in woodland pastures, but these trees are not very old.

American sycamore (*Platanus occidentalis*) is a conundrum. Today it is very common both along streams and rivers and on limestone uplands. Early narratives of the Inner Bluegrass do not mention this species, however, even though it is very easily recognized and was very useful because of its soft, workable wood. Early accounts tell us that it was present at Fort Boonesborough on the Kentucky River, and certainly along the Ohio River, but not, apparently, on the upland sites where it is found today. It is possible that the abundance of cane along creeks prevented sycamore, which is very intolerant of shade, from becoming established. Although sycamores over four hundred years old are known, large trees in the Bluegrass are always hollow, which prevents an accurate assessment of their age.

Osage-orange (*Maclura pomifera*) is another puzzling tree. It is one of the most abundant trees in the Bluegrass and Nashville Basin, but we don't really know its origin or how long it has been in our region. The native range of this tree is supposed to be a small region along the Red River in Oklahoma and Texas, and it would probably not be in the Bluegrass at all if it were not an enormously useful plant. For Indians, Osage-orange was the source of strong, flexible wood for making bows. One of its many names is bodark (from the French *bois d'arc*—wood for bows). Indians traded Osage-orange bows throughout the East, but they might also have traded fruits and seeds. Indians might have deliberately or inadvertently planted Osage-orange in the Bluegrass.

European settlers found Osage-orange wood very useful for wheel hubs and rims and for oxen yokes. But the great flourishing of Osage-orange in the Bluegrass came with the realization that tightly planted rows of thorny young Osage-orange made excellent fences. Before the introduction of barbed wire around 1870, farmers would plant Osage-orange trees a foot apart and weave the branches together, a process called plashing. We know that Meriwether Lewis sent Osage-orange cuttings to Thomas Jefferson in 1804, and many accounts of this tree date its eastward spread to that time. Osage-orange trees of great size and age can be found in the East, however. In Red Hill, Virginia, there is a huge Osage-orange that probably dates to the early 1700s. The immense Osage-orange at Fort Harrod, Kentucky, is probably of that age. It appears likely that Osage-orange was carried east by Indians long before the Lewis and Clark expedition.[10]

Today Osage-orange is one of the most abundant trees in the Bluegrass, where it is commonly known as hedge-apple. It is most common along hedgerows and fence lines, which perhaps reflects its original use as a living fence. It is also a common component of woodland pastures. Osage-orange reproduces well from stump sprouts and root sprouts, and an entire stand of this tree can originate as a single clone. We can speculate that some of these clones were here when the first settlers arrived, but we will probably never know.

Catalpa, or northern catalpa (*Catalpa speciosa*), is another common tree in our woodland pastures whose origin is complicated. Current range maps list catalpa as a native of the Mississippi River valley in western Kentucky. Gilbert Imlay, who lived in Fayette County, Kentucky, in 1783, listed catalpa among trees he had seen in the region, but it is hard to know how far west he had traveled. Catalpa was heavily promoted in the nineteenth century as a timber tree for railroad ties and later as an ornamental before it lost its popularity early in the twentieth century. Most of the horse farms of the Bluegrass have catalpa trees in yards, along fences, and, often, in woodland pastures, and it has established self-reproducing stands throughout the region. Catalpa is a dominant tree in some woodland pastures.[11]

Red mulberry (*Morus rubra*) was prized for its fruit by settlers and is mentioned in many early accounts of the Bluegrass. Today we find very few red mulberry trees, mostly in the understory of woods but occasionally in woodland pastures. Red mulberry is declining throughout its range and is critically endangered in Ontario. The reasons for its decline are not clear, but the introduced and invasive species white mulberry (*Morus alba*) has become much more abundant and interbreeds with the native red mulberry. Today in the Bluegrass we see many more white mulberry trees than red mulberry, especially in urban areas.

5

The Ingleside Oak

The Bluegrass and the Nashville Basin in 1779

A large, old bur oak stands by the side of Harrodsburg Road, one of the busiest roads in the Bluegrass, as thousands of cars whiz by every day. The tree was once on the edge of a huge estate known as Ingleside Manor, where cattle, sheep, and horses grazed within sight of downtown Lexington. Harrodsburg Road began as a buffalo trace and was one of the first paved roads in Lexington; the tree shaded the original buffalo trace. Over the years the road was widened and repaved, coming closer to the Ingleside Oak, but never doing it any real harm. Ingleside Manor is long gone, replaced by businesses and homes, and only a few trees remain of what was once a very large woodland pasture of venerable trees. The woodland pasture was here long before Ingleside Manor was established in the mid-1800s.

I like to think that Josiah Collins walked past this tree. We are not sure what route he took to arrive in Lexington in April 1779, but the buffalo trace would have been a convenient path that he must have trod at some point. The first houses that Collins helped build were only a mile from the Ingleside Oak.

On the day that Josiah Collins felled that first bur oak to build the blockhouse at the center of Lexington, the Bluegrass had barely been explored by white travelers and settlers. In their masterful book about the natural history of the Bluegrass, Mary Wharton and Roger Barbour summarize many of the narratives of early travelers and settlers, though there is

Figure 5.1. The Ingleside Oak.

Figure 5.2. Ingleside Manor in 1904. Photograph by Thomas Knight.

no substitute for reading the original accounts. It can be quite confusing to compare the stories of various travelers because they are so different from one another. The Bluegrass is large, encompassing more than 10,000 square miles. Travelers limited to rivers, horses, and their feet could not have seen it all. A traveler going up Elkhorn Creek on the north side might have seen a completely different landscape from the one a traveler going up the south side saw. Rather than conflicting with one another, the early accounts give a sense of the varied nature of our landscape.[1]

Many narratives from settlers and later from biologists have noted what a strange landscape this is. It was jarring and unexpected for early travelers to see open woodlands after months of travel in dense Appalachian forests. Neither forest, nor prairie, nor savanna, the Bluegrass was a source of wonder and delight. The botanist E. Lucy Braun later described the Bluegrass and Nashville Basin as the "most anomalous vegetation area of the eastern United States."[2]

Thomas Hanson went up the North Branch of the Elkhorn in July 1774 and wrote, "The land is so good that I cannot give it due praise. Its undergrowth is clover, peavine, cane. Its timber is honey locust, black walnut, sugar tree, hickory, iron-wood, hoop-wood [hackberry], mulberry, ash and elm and some oak." James Nourse, a year later and also on the Elkhorn, noted that "the further we went the richer the land, better though of the same sort of timber, the ash very large and high and large locusts of both sorts—some cherry—the growth of grass under amazing. . . . What would be called a fine swarth of grass in cultivated meadows and such was its appearance without end. . . . We passed several dry branches but no running water in our course southeast."[3]

Clearly from these and other accounts, the landscape encountered by these early explorers was quite astonishing and consisted in some places of open-grown trees shading grass, cane, and other herbaceous cover, not of dense forest. In his map of Kentucky of 1784 (see below), John Filson described the same area along the Elkhorn as "fine cane." Other areas were densely wooded, and these were the primary source of timber.

Descriptions of the Nashville Basin are a bit less clear, mainly because Nashville was established in 1779 in the more heavily forested Outer Nashville Basin along the Cumberland River. Settlement of the Inner Nashville Basin, which more closely resembles the Inner Bluegrass, came somewhat later with the settlement of Murfreesboro in 1811.

Braun and other biologists have puzzled over the origin of our unique landscape. To understand how it came into existence, we need to examine the roles of cane, bison, Indians, fire, and drought.

Cane

Josiah Collins got lost in the cane, perhaps more than once. He and a companion went in search of some missing horses, wandered into thick cane, and were unable to find their way back to Lexington. They finally found themselves several days later near Fort Boonesborough, many miles from their starting point, though it is not clear how much of that time they were in cane. Numerous accounts of cane speak of its vast extent and great height, especially along the upper banks of creeks.

Giant cane (*Arundinaria gigantea*) is our bamboo. Like most bamboos, it forms dense, tall stands along creeks and moist uplands. Cane is more sensitive to drought than most grasses, and cane and grasses probably shifted back and forth in wet and dry years. Some early accounts refer to great grasslands with trees without mentioning cane. Along creeks cane reached thirty feet in height, but it was shorter on upland sites. Cane was a favorite food of bison, which could browse down and trample a cane stand in short order.[4]

Dense stands of cane do not completely exclude trees, but they would slow their growth substantially. The venerable tree species vary in their tolerance of shade, but we know that they can hang on as saplings for quite some time in shady environments and then grow more quickly when they are released from the shade.

Both intense browsing by bison and drought would reduce the density and height of cane stands. The biology of cane also makes dense stands come and go. Cane, like other bamboos, has an unusual habit of mass flowering and death. All the cane in a stand, and sometimes over a region, can flower at once, bear seed, and then die. The following year, the stand begins to make a comeback from the seeds. The combination of bison foraging, drought, and intermittent flowering and death meant that cane would not completely shade out young trees, but instead would provide moderate shade and periodic exposure to full sun. Far from discouraging the establishment of trees, cane may have been an important part of their success. Although there are few stands of cane left in the Bluegrass, I have often seen seedlings of woodland pasture trees in cane stands. I found one stand of cane shading a group of small bur oaks, none of them taller than eighteen inches. A count of the rings on two of the trees showed that they had been growing slowly in the shade of the cane for nearly thirty years.

Figure 5.3. Giant cane beneath a bur oak in the morning sun.

Bison

Josiah Collins and other settlers were well aware of the presence and abundance of bison, also called buffalo, in the Bluegrass. American bison (*Bos bison bison*) were common in eastern North America before European settlement. In 1612 bison were first seen by East Coast explorers in the vicinity of what is now Washington, D.C.[5]

Figure 5.4. American bison in a Bluegrass pasture, blue ash in the background.

John Filson was impressed with the abundance of bison in the Bluegrass, noting that "the amazing herds of buffaloes which resort thither, by their size and number, fill the traveller with amazement and terror, especially when he beholds the prodigious roads they have made from all quarters, as if leading to some populous city; the vast space of land around these springs desolated as if by a ravaging enemy, and hills reduced to plains. . . . I have heard a hunter assert he saw above one thousand buffaloes at the Blue Licks at once; so numerous were they before the first settlers had wantonly sported away their lives."[6] Similar herds were present in the Nashville Basin. In 1770 there were "immense numbers of buffalo and other wild game. The country was crowded with them. Their bellowings sounded from the hills and forest."[7] In a mere thirty years, settlers reduced the bison population to a handful. By 1810 bison were rarely to be found east of the Mississippi.

American bison were not always found in Kentucky or Tennessee, and it is not known when they arrived here. Bison fossils at Big Bone Lick are from a different species, which was extinct by about 10,000 years ago. Bison remains are found only in a few mounds, middens, and other remains of pre-Columbian people in the Ohio River valley, but this scarcity may indicate the difficulty of hunting such huge animals before the introduction of the horse and the rifle. They were definitely here in very large numbers in the mid- to late 1700s.[8]

The American bison is the same species as the European bison, or wisent (*Bos bison bonasus*), but they differ in their feeding behavior. The American bison is a grazing animal, consuming grasses and foliage but not browsing on woody plants. The wisent is a mixed feeder, browsing on shrubs and trees as well as grazing grass. American bison and wisent are closely related to the yak (*Bos mutus*) and domestic cattle (*Bos taurus*), and all these species can interbreed. These distinctions in feeding behavior are important to our understanding of the role of large herbivores in our landscape compared with those in woodland pastures in Europe. In both cases, large herbivores are important in the maintenance of open woodlands, but bison and wisent have very different effects. American elk (also called wapiti, *Cervus canadensis*) and white-tailed deer (*Odocoileus virginianus*) were present, but evidently not in such large numbers as the bison. Though all early accounts describe the huge bison populations, elk and deer did not make the same impression. Nearly all early accounts of the diet of settlers indicate that bison was the main meat source, while deer and elk were a distant second. Large numbers of elk and deer munching on buds and stems can inhibit the growth and survival of trees, but they must not have been present in our woodland pastures in sufficient numbers to prevent the growth of trees.[8]

Although both bison and cattle are grazers, their behaviors are quite different. Cattle are slow, methodical grazers, bred to convert grass as efficiently as possible to milk and meat. Bison also graze in a slow, methodical fashion but are capable of amazing feats of agility, making leaps of over fourteen feet to cross cattle grates and running for great distances. Their natural behavior is to graze placidly for a while, and then to move with some haste from place to place. Bison herds will move twenty miles in a single night and hundreds of miles season to season. Unlike cattle, bison are athletic, long-distance wanderers that may forage heavily in one place but then quickly depart, perhaps not returning for many years.[9]

One important question about the vegetation of the Bluegrass and Nashville Basin is whether there was an abundance of thorny shrubs. In Europe, thickets of thorny shrubs play a critical role as nurse crops for oaks and other trees in woodland pastures, protecting them from browsing by wisent and red deer. Black locust and honeylocust were unlikely to play this role because they are fast-growing trees. American plum (*Prunus americana*) does form thickets in the Bluegrass and is slightly thorny. Several hawthorn species (*Crataegus* spp.) are common, especially in abandoned pastures and along fencerows, but they do not seem to form dense thickets that could protect young trees. I suspect, however, that the behavioral difference between the browsing wisent and the grazing bison may have made thorny trees and shrubs less important to the survival of trees in our woodland pastures.[10]

Indians

Josiah Collins settled in the Bluegrass, although he was often away for long periods serving in various militias. Later in his life he recounted his interactions with Indians, including skirmishes around Ft. Boonesborough, pitched battles along and north of the Ohio, and minor skirmishes around Harrodsburg. Yet he recalled very few adverse encounters with Indians around Lexington. Aside from seeing a few Indians and recalling several stolen horses and a few one-to-one encounters between Indians and setters, Collins recalled that he had no "commerce" with them, meaning no regular interactions. The Indians most often encountered were Shawnee, who had villages along the Ohio, Scioto, and Kanawha rivers, but not in the Inner Bluegrass.[11]

Eskippakithiki was a Shawnee farm village on the border between the Outer Bluegrass and the Knobs. It was the only known permanent Indian settlement in the Bluegrass. Eskippakithiki was in what is now called the Indian Old Fields, a remarkably level area of about 3,500 acres along several creeks and near the Kentucky River. One story about the origin of the name Kentucky is that it was the Virginia Algonquian word *Kenta-ke*, meaning a level place of fields, which referred to this location.[12]

Collins and others recalled finding winter hunting camps scattered around the Bluegrass. Today farm owners often find evidence of winter hunting; arrowheads and spear points turn up in plowed fields. Permanent villages were not present in the Bluegrass, except for Eskippakithiki, when the first Europeans arrived. Cherokees came into the Bluegrass on occasion, but apparently not in large numbers.

Why were there so few Indians in a region that seemed to have everything they would need? Many historians describe the Bluegrass as an area of disputed land ownership among tribes. While social factors are important,

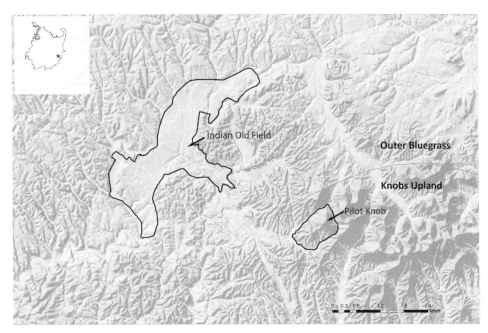

Figure 5.5. Map of the Indian Old Field, the site of Eskippakithiki. The inset shows the location of the Indian Old Field in Clark County.

Figure 5.6. The Indian Old Field today.

it is at least as likely that environmental factors, including disease, drought, and bison, combined to prevent Indian settlements except around the edges of the Bluegrass.

European diseases such as smallpox, flu, and cold viruses arrived in advance of European settlers. There is no question that disease dramatically reduced Indian populations throughout the region, but there were still many Indian villages north, east, and south of the Bluegrass well after European settlement.

Drought probably played a very large role in where Indians chose to settle. As the creeks and springs of the Bluegrass dried up, the Shawnee may have been compelled to retreat to the larger rivers to the north and east, and the Cherokee to the south. The period from 1746 to 1755 was profoundly dry; normal precipitation fell in only two of those years. This alone may have caused the abandonment of Eskippakithiki in 1754.

As anyone who lives here can attest, the Inner Bluegrass and Inner Nashville Basin become quite dry in summer and autumn, even in otherwise wet years. Most streams are seasonal, and there are usually several months during which surface water is hard to find. This droughtiness is due to the karst topography, which allows rain and melting snow to quickly trickle down into the bedrock. The seasonal lack of water may have discouraged permanent settlements in either area.

Bison were probably a major factor in Indian settlements. Although we commonly view bison as a source of food and skins for Indians, large herds of bison wandering through the Bluegrass may have discouraged permanent habitation. Early European accounts tell of bison knocking down cabins and consuming corn crops, and of people seeking shelter behind stockades or fleeing when the bison were passing through. One settler noted, "Buffaloes used to be passing by Lexington every day and sometimes all day long. Virginians used to come out and spend the winter [in the Bluegrass], and go back again in the spring before dangerous times commenced." From the context of this narrative, this was not a reference to Indians, who were also present largely in winter, but to bison.[13]

Without rifles and horses, Indians may not have been able to withstand large herds of foraging bison. Springs and other reliable water sources, in the absence of topographical features to discourage bison, may have been too dangerous. In drought years, bison would have wandered even more extensively.

Eskippakithiki was very level, but it was surrounded by steeper terrain and rivers that provided protection from bison. I have found it very difficult to walk from the level Indian Old Fields, ford the creeks, and ascend steep banks and hills. It may have been the only location in the Bluegrass with reliable water and protection from both bison and tribal conflict. The close proximity to the heavily forested and hilly Knobs, only a mile to the east, may have also provided protection from the more powerful Cherokees who wandered through the area.

The situation in the Nashville Basin was quite different. The Cumberland River flows through the northern Basin and provides water even in drought years. Cherokees had villages throughout the area, though perhaps not in the Inner Nashville Basin, away from the Cumberland River.

Historians recount extensive battles between settlers and Cherokees for several years after the establishment of Nashville in 1779.

The Fort Ancient and other prehistoric cultures were present in the region for thousands of years, perhaps as late as 1750. Like the Shawnee, they seem not to have lived in most of the Bluegrass; their settlements were generally on or near the Ohio River, although there are some mounds and other evidence of habitation scattered throughout the Bluegrass. Eskippakithiki was probably a Fort Ancient village before it was Shawnee. The Fort Ancient people may have become the Shawnee, although this is not known.

For our understanding of the formation of the Bluegrass, the critical lesson from Eskippakithiki is that permanent settlements in the Bluegrass may have been discouraged by the huge numbers of wandering bison and the frequent occurrence of drought, especially in the years just before European settlement.

Fire

Fire plays an important role in the creation of savannas, prairies, and openings. Without fire, savannas and tallgrass prairies are quickly converted to forest. Indians used fire as a management tool throughout the East, burning

Figure 5.7. A grass fire in central Kentucky. Fires, started mostly by people, are common in grasslands but rare in the Bluegrass except during drought. Trees in the Bluegrass, even very old ones, do not bear fire scars.

off areas to make clearings to farm and to increase game populations. Yet there is strong evidence that fire was not a factor in establishing the mosaic of vegetation in the Bluegrass.

Every tree maintains a personal diary of its life in the form of annual rings. Drought makes rings narrow, abnormal frost makes characteristic marks in the rings, and fire leaves scars. Yet in all the cores taken by forest biologists, and in all the wood we have seen from dead or downed trees, we don't see fire scars in Bluegrass trees. In contrast, old trees in oak savannas and in areas where Indians used fire as a management tool typically have fire scars. Lightning scars, distinct from fire scars, are very common in old trees, especially blue ash, but lightning strikes do not cause substantial fires. I have seen two blue ash trees being struck by lightning. In both cases, the strikes created great puffs of steam and smoke, but no fire that spread to the ground, and little charring of the wood.

In the Bluegrass, pioneer accounts rarely mention fire until after permanent settlement, when white settlers began using fire to drive game. It is possible to set grass fires in the Bluegrass, but they tend to burn themselves out and do not become raging wildfires.

Drought

One of the paradoxes of life in the Bluegrass and Nashville Basin is that we get plenty of rainfall but are still prone to drought. That is because of the limestone karst topography, in which soil lies on top of very porous, cracked, and fissured bedrock. Rainfall can quickly percolate down through the soil, into the bedrock and into underground water courses. Most of the smaller streams dry up in the late summer and early fall, and in the majority of years grass is dry and brown from August to October.

To figure out the role of drought in the past, scientists have turned to the detailed mathematical analysis of tree rings. Dendrochronology—the telling of time with trees—allows us to evaluate the influence of fire and drought in the past. Trees are very sensitive to drought, and each drought leaves a record in their annual rings. A very narrow ring in many trees in a region usually indicates drought during the growing season, while a wide ring indicates average or above average rainfall. Correlating growth rings among large numbers of trees has allowed dendrochronologists to calculate the frequency of drought for most regions of North America for more than 1,500 years in the past. Analysis of the tree ring data for our region shows that mega-droughts lasting more than a decade have occurred several

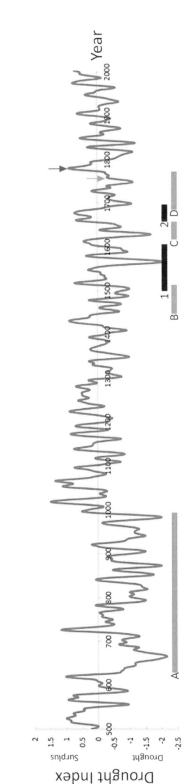

Figure 5.8. The Palmer Drought Severity Index, twenty-year running averages, obtained by tree-ring analysis, from 500 to 2003. Drought index values below zero are growing-season drought periods. Prolonged drought periods are indicated by letters (A, B, C, D); years when tree ring counts indicate the establishment of trees are indicated by numbers (1, 2). The left arrow is 1754, the year of abandonment of Eskipakithiki, and the right arrow is 1779, the year Lexington was settled. Prolonged drought periods were defined as periods when more than 70 percent of years in the period were in drought. A indicates a 368-year drought period beginning in 625; B indicates a 67-year drought period beginning in 1447; C indicates a 38-year drought period beginning in 1619; D indicates an 87-year drought period beginning in 1687. (See the sources cited in note 14 to this chapter.)

times, including a series of droughts just before European settlement, and a remarkable period of over 350 years of recurring drought 1,000 years ago. This does not mean that every year during those mega-droughts was dry, but that there were long periods during which most years were dry. The entire eastern United States was in serious drought from 1772 to 1775, and this caused an opening up of forests as drought-sensitive trees died.[14]

Because of our karst topography, these mega-droughts would have had much greater effects on life in the Bluegrass and the Nashville Basin than in other regions. These multiyear droughts are probably the most important factor in explaining our unusual landscape of venerable trees.

It appears likely that drought, karst, and bison were the three main factors that created the conditions under which woodland pastures developed with little human influence. In the next chapter, we will compare our woodland pastures with those of Europe and see how the landscape of 1779 was formed.

Settlement and Tree Growth

There are other explanations for the trees in our midst. Ryan McEwan and Brian McCarthy analyzed tree growth in several woodland pastures and found that there was a long period of suppressed growth in trees, followed by their release at around the time of the establishment of the first farms in the region. They interpreted this to mean that the woodland pasture trees became established in forests, and that settlers thinned out the forests, creating the woodland pasture landscape that we see today.[15]

It is certainly true that the majority of ancient trees had a prolonged period of slow or suppressed growth early in their lives. It seems very unlikely, however, that settlers would have chosen to thin the forests and leave only the bur oak, chinkapin oak, Shumard oak, blue ash, and kingnut trees behind. There are many areas of the eastern United States where farmers cleared forests to establish pastures, but they did not create the woodland pastures that we have in the Bluegrass and Nashville Basin. If partial clearance of land to create shaded woodland pastures was a common farming practice, we would expect to find remnant woodland pastures throughout the East, but we do not. As Braun pointed out, this is the most unusual habitat in eastern North America, and it is unlikely to have been uniquely created by management practices.

The idea that this area was heavily forested and that early settlers cleared out some of the trees, selectively leaving behind the particular tree species we find today, is also not entirely consistent with the historical record. It is certainly true that there were forests here in some areas, but many early descriptions are quite convincing that there were open-grown woodlands containing grass and cane.

The abundance of bison is perhaps the most compelling evidence that the Bluegrass was a mosaic of woodland pastures, meadows of cane and grass, and forests. Neither the American bison nor the European wisent is a woodland animal. Each requires open woodlands or grasslands. Had the Bluegrass consisted primarily of closed-canopy forests, the bison would not have been here in significant numbers. The abundance of wisent and other large herbivores in Europe has been accepted as evidence that the European landscape also consisted of woodland pastures before widespread settlement and farming.

As we have seen, our woodland pasture trees can grow in the shade of cane and tall grass. Cane was as tall as thirty feet and very dense, as shown by the number of settlers who became lost in it. It is quite likely that the suppressed growth of the woodland pasture trees observed by McEwan and McCarthy, and by others, was a result of their growth for long periods in the shade of a mixture of grass, cane, and trees. Woodland pastures differ widely in their tree density, and it is possible that trees growing in the shade of a woodland pasture would be suppressed for a long time, whether shaded by cane or by older trees.

These are not mutually exclusive ideas. The Bluegrass was a mosaic of different vegetation types, and scientists have sampled only a small number of trees for detailed age analysis. There was probably a continuum in tree density from open grasslands to woodland pastures to forests. Even today some woodland pastures are quite a bit denser than others. Further research is warranted, but we may never know the whole story.

6

The Woodland Pasture

Woodland Pastures of Europe

Woodland pastures are a common landscape feature in Europe, where animals graze under the shade of trees, often very old trees, and where fire is not important. The tree cover may vary from a few scattered trees to dense swards, but it is never enough to shade out the grasses and herbs that form the ground cover. The open woodlands of the Bluegrass and Nashville Basin bear a remarkable resemblance to the woodland pastures of Europe. Woodland pastures are generally called wood pastures in Europe, but they are the same thing.[1]

Oliver Rackham, the renowned English botanist, was the first to recognize that the woodland pastures common in the landscapes of the United Kingdom are not recent, but date from medieval times or earlier. Within the U.K. woodland pasture landscape are many ancient trees, including oaks over one thousand years old. Because of the work of Rackham and his colleagues, there is now a concerted effort to create a sustainable woodland pasture management system so that these important landscapes survive the next one thousand years.[2]

Frans Vera is a Dutch natural resource scientist who has done extensive work on woodland pastures throughout Europe. He has proposed that the open woodland pasture landscape so common in Europe is maintained by human activity but was created by natural processes before the invention of modern agriculture. He has shown that grazing and browsing animals played an important role in creating the woodland pasture landscape, and farmers then took advantage of the existing landscape to create sustainable pasture grazing systems. The grazing and browsing animals included wisent, or European bison (*Bos bison bonasus*); auroch (*Bos primigenius*),

Figure 6.1. A woodland pasture in Kentucky.

Figure 6.2. A woodland pasture in Romania. Photograph by Tibor Hartel; used by permission.

Figure 6.3. A woodland pasture in England. Photograph by Julian Hight; used by permission.

Figure 6.4. A woodland pasture in Russia. Painting by Ivan Shiskin, 1887, collection of the State Tretyakov Gallery, Moscow, used by permission.

the wild ancestor of modern cattle;[3] European moose (*Alces alces*);[4] red deer (*Cervus elaphus*);[5] and horses (*Equus caballus*).[6] Many of these animals feed mostly by browsing on shrubs and trees, rather than grazing. Wisent are both grazers and browsers, consuming mostly grass and herbs but also some buds and twigs of shrubs and trees. In contrast, the American bison, like modern cattle, is almost exclusively a grazing animal, eating grass and herbs.[7]

Some biologists have argued that wisent were creatures of the forest in the Holocene epoch (11,700 years ago until the present), not primarily of open woodlands. Recent research on wisent of the Holocene, however, shows that their diet was a mixture of grass and woody plants, whereas moose and red deer ate mostly woody plants and auroch were grazers. Moose and red deer were primarily forest animals, aurochs were found in open pastures, and wisent were in woodland pastures. This is consistent with the conclusions of Rackham and Vera that large herbivores, including auroch, horses, and wisent, were instrumental in maintaining the open wood pasture landscape.[8]

Today the woodland pastures of Europe are threatened by age, land conversion and development, and a lack of natural regeneration, just as we see in the Bluegrass today. As Europe's population becomes more urban, and there is more underused land, the opportunities for a return to woodland pasture landscapes are increasing. Frans Vera is now leading the Oostvaardersplassen project in the Netherlands and other programs to maintain existing woodland pastures and to restore woodland pastures to areas where they have been long gone.

In Romania, Tibor Hartel and other scientists are leading efforts to understand and maintain the existing woodland pasture landscape and to "rewild" the landscape by reintroducing some of the original animals. In 2014 wisent were reintroduced to this landscape after an absence of over 250 years. One major difference between European woodland pastures and ours is the browsing behavior of wisent in contrast to the grazing behavior of American bison. Hartel and Vera have found that thorny shrubs are necessary in European woodland pastures to protect young oaks and other trees from browsing. Thorny shrubs may be less important in Bluegrass landscapes.

It is important to understand the foraging behavior of the animals that create woodland pasture landscapes, because the differences are important to our understanding of how to manage them. Bison are gone from

Figure 6.5. A woodland pasture in Kentucky, 1904. Photograph by Thomas Knight.

our landscape and are not coming back in any great numbers. If we can mimic their behavior of intense grazing followed by long periods of absence, then we may be able to re-create the conditions under which venerable trees reproduce and thrive. As we will see in the next chapter, there is some evidence that this strategy is effective.

In the Bluegrass, it was common to refer to this landscape as woodland pasture. Thomas Knight's vivid photographs portray farms in central Kentucky at the beginning of the twentieth century and clearly show extensive woodland pastures on most of the famous horse farms and estates. Early descriptions of the Bluegrass and Nashville Basin referred to open woodlands, woods with cane and grass, open-grown trees, woodland meadows, and woodland pasture.

The woodland pastures of the Bluegrass and Nashville Basin are sometimes called oak savannas (or savannahs). *Savanna* is a term used by biologists to designate landscapes that form a transition between grasslands and forests. In North America the term *savanna* is used for open oak woodlands that are transitional between the tallgrass prairie of the Midwest and the deciduous forests of the East (see figure 4.1). The true savanna of the

Midwest is an ecosystem created and maintained by frequent fire and low rainfall. Of the venerable tree species in our woodland pastures, only bur oak is characteristic of midwestern savanna. The definitive book on American savannas, by Anderson, Fralish, and Baskin, specifically excludes the woodland pastures of our region from consideration as savanna.[9]

The distinction between savanna and woodland pasture is important, because the way we describe landscapes influences how we think about them. If we call them savannas, then we will want to manage them like the oak savannas of the Midwest, even though we know that fire was not important here. Indeed, the Kentucky Department of Fish and Wildlife has been using fire to manage Griffith Woods, a magnificent woodland pasture, in the mistaken belief that it is applying appropriate management techniques.

If we manage our venerable trees as woodland pastures, then we can use the rich history of European wood pasture management for inspiration. We can learn a great deal about management of our woodland pastures from the European experience, but far less from understanding management of Midwest oak savanna.

The Woodland Pasture Scenario—Bison, Drought, and Trees

We know that woodland pastures were part of the landscape of the Bluegrass and Nashville Basin landscapes at the time the first European settlers arrived here. We also know that this landscape was unique, astonishing to the first explorers, and not found anywhere else. Two factors can account for the existence of this landscape: a series of multiyear mega-droughts made worse by karst drainage, and the presence of large numbers of American bison. Together, they created a mosaic landscape that included the extensive woodland pastures that we see today. What follows is a scenario for the structure of the Bluegrass in 1779 that seeks to explain the origin of today's woodland pasture landscape.

The banks of the larger rivers, including the Kentucky, Licking, Salt, and Ohio in the Bluegrass and the Cumberland in the Nashville Basin, were densely forested with hardwood and redcedar. Steeper slopes, especially in the Hills of the Bluegrass and the Outer Nashville Basin, were also forested. The smaller creeks, such as the Elkhorn and its tributaries, were lined with cane in dense stands (canebrakes) that prevented trees from growing. In the Inner and Outer Bluegrass, as well as the Inner Nashville Basin, upland sites away from the creeks supported cane of lower stature, along with grasses

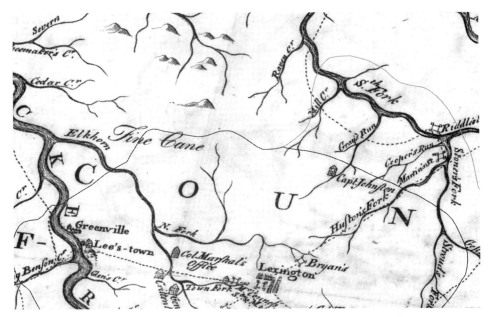

Figure 6.6. A detail of the Filson map of 1784. This map was digitally fitted to a modern map of the Bluegrass, so that figures 6.6 and 6.7 can be compared. Many accounts of the area along the Elkhorn from Leestown (Frankfort) to Lexington exist, including Filson's description of "Fine Cane" there.

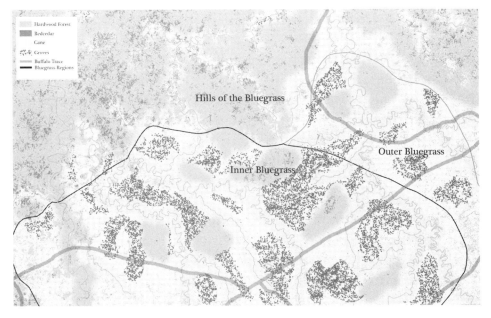

Figure 6.7. A digital reconstruction of the Bluegrass in 1779, in the same area as shown in figure 6.6. Dense, tall cane along the creek banks probably prevented woodland pasture trees from becoming established. Uplands were a mosaic of forest, woodland pasture, and meadows of grass and cane.

such as wild rye. Buffalo traces crisscrossed the landscape, fording the rivers at low points and creating wallows and stomping grounds. Filson's map of 1784 shows the creeks, buffalo traces, and "fine cane" around the Elkhorn, and other early accounts tell of the mosaic of forest, grasslands, and woodland pasture.

The Bluegrass and the Nashville Basin provided the huge bison herds with everything they needed—abundant cane and grass, springs and rivers for water, and natural mineral licks to provide important elements such as sodium. These resources were unevenly distributed: the salt licks were around the edges of the landscape, and the smaller streams dried up in summer. Bison wandered the landscape, moving in herds among forage areas and to and from salt licks and reliable water sources.

Groves of venerable trees formed a mosaic on uplands mixed with more open meadows and closed woodlands. The mosaic was maintained by the habits of bison. Bison in large herds would intensely graze an area, reducing cane and grasses to stubble. Then, as the creeks dried up in summer and autumn, bison herds would migrate along the traces to salt licks and permanent water sources. They may not have returned to an area again for several years or even decades.

In the spring, following the bison's departure, dense stands of tree seedlings germinated from seeds deposited in the previous fall or stored in the soil from earlier years. These seedlings grew rapidly for a year or two, and then more slowly as the cane and grasses overtopped them. There was sufficient light for the seedlings to survive and grow slowly, even in the shade of the cane. When the bison returned, they would again graze down the cane and grass, releasing the trees. If the bison returned too soon, then they might trample the young trees. If they did not return for many years, then the trees would be robust enough to withstand the trampling. Soon the groves of trees would be taller than the cane and grasses, and the resulting stands could survive for centuries. These were woodland pastures, created and maintained by the natural behavior of bison.

Drought played a critical role in reducing competition from grass and cane, and especially from less drought-tolerant trees. The karst topography strengthened the effects of drought. Once established, the trees had deeper root systems than cane or grass and could survive the drought years. Drought also stimulates trees to produce abundant seed crops, and it is common to see heavy tree reproduction right after a drought. There were several periods of prolonged drought in the years just before settlement. It appears that some of the trees here today began growing right at the ends of these periods (see figure 5.8).

The prolonged mega-drought beginning in the year 625 could have been the most important factor in creating the Bluegrass mosaic in which venerable trees could thrive. Before that time, the landscape may have been entirely forested. As Lucy Braun and others have noted, an area with the high soil fertility and high rainfall of the Bluegrass and Nashville Basin should be forested. But a very long drought period, along with drought-prone soils, could have created the landscape of open woodlands, cane, and grass. The arrival of bison some time during or after that period could have maintained the landscape in this condition until 1779.[10]

Then the settlers came. Indians largely left the area by 1754. The abandonment of Eskippakithiki in 1754 followed twenty years of severe drought, and there were several more significant droughts between 1754 and 1775. Neil Pederson and his colleagues have recently found, through tree-ring analysis, that there was an intense drought throughout eastern North America from 1772 until 1775. The period from 1775 to 1779 was wetter than average and would have provided the first European farmers with excellent growing conditions.[11]

Early accounts of the Bluegrass sang the praises of open lands and vistas unseen in the wilderness to the east. The first settlers may have been subsistence farmers, but they were quickly replaced by wealthier landowners who acquired large tracts of land. Though bison were intermittent grazers, cattle and sheep were not permitted to wander, and they rapidly consumed the native ground cover. The cane and native grasses were quickly grazed into oblivion and replaced with nonnative forage.

The groves of trees were kept, though some were thinned. Open-grown trees have lower value for timber than trees growing in nearby forests. There was probably a gradient between open-grown trees of the woodland pastures and higher-density forests on some sites. Thinning would have reduced the tree density while leaving the open-grown appearance of the woodland pasture. Some of the trees that were retained, especially sugar maples, were abusively managed and died.

The open-grown trees had tremendous value as woodland pasture, shading cattle and sheep and, later, horses. Landowners also wanted to keep groves of shade trees around their homes, many of which were impressive mansions.

Domestic livestock and fences fundamentally altered the environment for trees. Though the trees that existed when the pastures were enclosed continued growing, seedlings could not reproduce. Domestic livestock grazed continually, and the native cane and grasses were replaced with dense grasses and herbs more suited to continuous grazing. The established trees continued to grow and thrive, but they gradually succumbed to age, soil compaction, and development and were unable to reproduce. As the old trees die, they either are not replaced or are replaced with ornamental trees poorly adapted to the woodland pasture environment.

Today's landscape is the direct descendant of the mosaic created by karst, drought, and bison. The bison, cane, and many of the forests are gone. The woodland pastures remain but are fading away.

Woodland Pastures after 1779

We have seen that woodland pastures were an important component of the landscape of the Bluegrass and Inner Nashville Basin at the time of settlement, around 1779. We also know that today the trees of the woodland pasture are not regenerating naturally because of constant grazing and mowing and the use of heavy farm equipment.

Yet to the careful observer, it appears that there are some woodland pasture trees that are not old enough to have been here in 1779. We have counted annual rings in a few large trees and found that some were one hundred to two hundred years old. Many blue ash trees, which are the most abundant trees in our woodland pastures, do not appear large enough to be over two hundred years old. We have also found a few chinkapin and bur oaks that do not appear to be much older than one hundred years, although judging tree age by size is fraught with peril. Our sample size is very limited, and we don't know the real age of most of these trees.

If it is true that there are some woodland pasture trees that are less than two hundred years old, and if it is also true that there are almost none that are less than one hundred years old, then we have to conclude that there was a period after settlement when natural regeneration continued. We can imagine that early pasture management was of low enough intensity that trees would be left alone by livestock long enough to survive. If livestock numbers per acre were low, and there was little or no mowing, then the opportunity for natural regeneration of woodland pasture species may have persisted in some locations for 150 years or so.

The International Harvester Farmall changed everything. The introduction in 1925 of the first modern, gasoline-powered, low-cost tractor, quickly followed by designs from John Deere and other manufacturers, revolutionized pasture management. Horse-drawn equipment was too slow, and steam tractors were too heavy and cumbersome. The gas-powered tractor dramatically improved the lives and productivity of farmers, and by the start of World War II there were over 2 million tractors in use on American farms. I suspect, but do not know for certain, that the intensification of agriculture allowed by the gas-powered tractor marked the end of natural regeneration of woodland pastures.[12]

7

The Mother Tree

Reproduction of Venerable Trees

The mother tree stands amid a flock of her children in Griffith Woods. Over ninety feet tall, she hovers over hundreds of young trees arrayed around her. The young trees are of many ages, from small seedlings to robust saplings more than thirty feet tall.

The mother tree is a kingnut. The name *kingnut* is not only a reference to the size of the nut but to its importance. For Shawnee and earlier people as well as for European and African settlers, kingnut was a reliable source of delicious nuts, high in fat and protein. The nuts are a bit hard to crack, but the reward is large. It is that deliciousness that enlists the help of animals to move her seeds out of the shade of her own canopy and provides her progeny with a chance at survival.

I call her the mother tree, but she is also a he and therefore also a father tree. Sex in trees is very complicated, perhaps more complicated than in animals, including humans. Depending on the species, a tree may be all male or all female, may be both male and female in the same tree, may be male in the early spring and female a bit later, or may start life as male and later become female.

Trees would have trouble finding suitable mates if it weren't for the dating services provided by wind and animals. In the tropics, trees are pollinated by animals—insects, birds, and bats. In the boreal forest, trees are pollinated by wind. In between, in the temperate forests, some trees are

Figure 7.1. The mother tree surrounded by her progeny.

pollinated by wind and some by insects, and a few by both. The mother tree relies on wind. Bees and flies may visit her flowers in spring, but only to steal pollen, not to buzz from tree to tree. Oaks are exclusively wind pollinated. Blue ash is mostly wind pollinated, though the flowers are visited by insects that may do some pollination.

Figure 7.2. Male flowers of the mother tree. Male flowers develop first and shed pollen before the female flowers develop.

Wind is not a reliable matchmaker. The mother tree has to make a huge amount of pollen, perhaps as many as 10 billion grains, only to toss them into the air. Pollen has to travel far because the distance to a receptive female flower may be great. The mother tree's pollen may travel hundreds of miles. In the spring, the atmosphere is filled with tree pollen.

Making a huge amount of pollen and scattering it to the wind may be good for the mother tree and her kin, but it is certainly hard on people in the region. Hickory pollen is an allergen, and many people in the Bluegrass suffer when the hickories are in bloom. Hickory pollen has to be small enough to carry far on the wind, and it is therefore small enough to be inhaled into our lungs. Pollen is coated in complex proteins, which help the pollen grain communicate with the surface of the female flower. Unfortunately for us, these proteins cause allergic reactions in many people.

If the pollen of the mother tree were to land on the female flower of the same tree, then the female flower would reject the advances of the pollen. Trees, except in the tropics, do not mate with themselves. Instead, they seek mates among other trees of the same, or closely related, species.

Using wind for fertilization presents the mother tree with a problem—how to create vast amounts of pollen to scatter to the wind in search of a mate, but not to drop all that pollen on her own female flowers. Timing is the solution. The male flowers, borne high on the tree to catch the wind, develop before the female flowers. The female flowers develop a few days later and catch pollen from other trees.

The surface of the female flower is sticky when she is ready to mate. Pollen from many species will land on that surface, but only pollen with the right protein will be allowed to fertilize the egg. From that mating comes the future kingnut tree, an embryo inside a shell inside a husk.

Once that fruit is ripe in the fall, more help is needed to get around. Gravity can carry seeds down from the tree, but a tree that can only reproduce right under the mother tree will not be very successful. Trees use wind, water, and animals to get around, and each tree species is adapted for one or more ways of getting around. Trees that disperse their seeds in the wind commonly have light, often winged seeds. Trees that depend on animals may create a sticky or hairy surface to attach to fur and feathers, or they may offer a food reward for animals to carry the seeds away.

Blue ash fruits, called samaras, are light and have an aerodynamic, curved wing. Blue ash waits to release its seeds until late fall and even winter, when the leaves are gone and the wind can fully catch the samaras and carry them a long distance.

The mother tree will not allow animals to take her precious children before the seeds are ripe, and she keeps those seeds bitter and unpleasant until the embryos are ready. Some trees advertise when their fruits are ready by changing color, becoming bright red or yellow instead of green. Kingnut fruits drop to the ground when they are ripe.

When the seeds are ripe, the mother tree makes a layer of cork in the fruit stem, and, as that corky layer breaks, the fruit drops to the ground. Gravity is not a great help to a kingnut tree. Blue ash, with its winged fruit that flies in the wind, can defy gravity and spread rather far without any other help. But these huge kingnuts drop with a thunk to the ground.

And then the animals come, squirrels mainly. The nuts are too large to interest most birds, though I have seen turkeys work at a kingnut. To a squirrel, a kingnut is a perfect food, rich in fat and protein, but opening it is a lot of work. The husk of the kingnut falls off when it is ripe, but squirrels can be impatient, and you can watch them work hard to husk a kingnut. And then they bury the nuts, as squirrels will do. And lose them. The genius

Figure 7.3. Fruit of the mother tree. Each thick-walled fruit contains a single seed, which contains a single embryo.

of the hickory-squirrel relationship is that squirrels are forgetful. If they remembered where every seed was buried, then the hickory trees would never have a chance to grow. If they forgot where every seed was buried, then the squirrels would not thrive.

Instead, the squirrels bury a nut that is a perfect packet of food both for the squirrel and for the future tree seedling. The fats, proteins, carbohydrates, and antioxidants provide complete nutrition for the squirrel, but also for the young tree as it begins to germinate the following spring.

It is tempting to think that this is an example of coevolution, that the tree evolved its large, nutritious fruit to benefit the squirrel, and the squirrel evolved its absent-minded burial to benefit the tree. And this may be true, though it is impossible to prove. There have been other players in the evolution of the kingnut.

Hickories became common in North America during the last 60 million years. The hickories evolved at the same time as the population of mammal species in North America exploded following the death of the dinosaurs, which eventually led to the rise of the giant mammals of the Pleistocene about 1.5 million years ago. Many mammals, from giant rodents and huge, piglike peccaries to ground sloths the size of a grizzly bear, evolved, thrived, and became extinct when modern hickories were growing. Biologists don't know much about the diets of these ancient mammals, but it is easy to think that they would have been consumers of kingnuts. Today, squirrels cache kingnut seeds, but it is not hard to imagine that a much larger animal may have swallowed the fruits whole, absorbing nutrients from the thick husk and excreting the seeds with all their nutrients in its feces. An animal would have had to be quite large to swallow whole kingnut fruits. I like to imagine giant ground sloths and shovel-nosed peccaries munching placidly on kingnut fruits, a few seeds surviving passage through their guts.[1]

There is more than one reason for a fruit to be very large. Upon germination, a kingnut can rapidly put down a very deep root system, using the nutrients it has stored. This makes the kingnut very resistant to drought even when it is quite young. It is also more resilient to herbivores, because most of its nutrients are stored in the root system. If the shoot is nipped off by a hungry elk or deer, it can regenerate a new shoot quickly. It is interesting that two of our venerable tree species, kingnut and bur oak, have the largest fruit of any of their relatives.

The mother tree is part of a very old woodland pasture at Griffith Woods Wildlife Management Area in Harrison County, Kentucky, in the Outer Bluegrass. The woodland pasture at Griffith Woods includes some of the oldest and largest trees in the Bluegrass. Griffith Woods was formerly a well-maintained livestock farm whose owners were very careful to keep the original woodland pasture scenery of their farm.

Throughout the region, it is difficult to find many places where venerable tree species are reproducing well. In fact, without our intervention, these trees will cease to exist in the Bluegrass and Nashville Basin. In surveys of dozens of woodland pastures, I have found adequate reproduction in only a handful of them. Why, then, is the mother tree so successful in producing young trees?

Cattle have not grazed on this woodland pasture in several decades, and this is probably why the mother tree has been able to reproduce. In the complete absence of cattle, this would become a forest, and the kingnut trees would be shaded out. If cattle were present constantly, then their foraging and trampling would prevent the seeds from germinating. It is the presence

Figure 7.4. Saplings, progeny of the mother tree, surround her.

of intense grazing followed by a long absence, mimicking the behavior of presettlement bison, that has allowed the mother tree to reproduce.

And her reproduction is prodigious: immediately around the mother tree are over 260 kingnut trees ranging from tiny seedlings to fifty-foot-tall saplings. There are also older kingnuts that may represent a previous pulse of reproduction. And these trees are not alone. There are patches of young blue ash in Griffith Woods and some young chinkapin oaks. Along with them are the other characteristic species, such as honeylocust, American elm, and white ash.

All the Bluegrass woodland pasture species are reproducing at Griffith Woods because the property has not been grazed for a few years. Grazing in some areas stopped in 2002, when the property was acquired by the Nature Conservancy, but other parts of the property were probably not grazed for about thirty years, the age of some of the kingnut saplings. The lack of grazing allowed the establishment of the young kingnuts and other woodland pasture species. Now, though, the absence of grazing is allowing Griffith Woods to become a forest that will gradually replace the woodland pasture, and the venerable tree species will be lost.

There is a way to ensure that places like Griffith Woods continue to exist as woodland pasture with adequate reproduction, and that is to introduce periodic grazing. The lesson of European wood pasture management is that some level of grazing is necessary. This does not mean we need to reintroduce bison, but, rather, that we need to allow cattle to graze intensively for short periods, followed by long periods without any grazing.

This approach requires a change in how we think about managing nature preserves. We tend to regard nature as being separate from the activities of man. But experience with managing bison in western prairies, and with managing wood pastures in Europe, shows that livestock, wildlife, and nature preserves can coexist. A woodland pasture preserve adjacent to an operating cattle farm provides the opportunity to allow nature to take its course in the preserve, by using the periodic introduction of cattle to prevent the woodland pasture from becoming a forest.

Jane Julian's property is both farm and nature preserve in Franklin County, Kentucky. It contains a large grove of venerable trees with a substantial amount of blue ash, chinkapin oak, and kingnut reproduction. Julian told me that she keeps the cattle out of the nature preserve most of the time, but lets them in every few years to "clean the place up." The result of this intermittent grazing is a fine stand of ancient trees shading a vigorous population of young trees.

In Woodford County, the Chandler Farm is a treasure trove of venerable trees, part of an extensive complex of groves covering many farms that we have only begun to explore and map. Some years ago, Toss Chandler put a tree pen, a fenced-in area for trees, around a large, declining bur oak. The tree died soon after, but today the pen contains a vigorous pure stand of bur oaks. Cattle continue to graze in the rest of the pasture, but the tree pen has allowed the old bur oak to reproduce.

The Kentucky River has carved a deep channel into the Bluegrass. In the Inner Bluegrass of Woodford County, the Huskisson Farm consists of upland pastures leading gently over a slope into forests along the Kentucky River. Near the upper edge of the woods is a dense stand of blue ash, from tiny yearlings to pole-size teenagers. This patch of woods shows clear signs of heavy grazing in the recent past. Previous landowners allowed their cattle to graze all the way down to the river, but the heaviest grazing was next to the pastures at the top of the hill. The combination of heavy grazing followed by the removal of livestock created the conditions for blue ash to seed successfully.

Figure 7.5. A bur oak acorn. The enormous fruit of the bur oak stores a large amount of carbohydrate, protein, and fat for the new seedling.

In a new neighborhood development in Lexington, a different disturbance has favored bur oaks. Property on Willman Way was formerly a woodland pasture with a creek running through it. The creek became part of a detention basin, designed to soak up water and prevent flooding during heavy rain. The detention basin was graded with bulldozers, but a number of very large, old trees were left. Three years later, a stand of young bur oaks has come up where the soil had been disturbed. The bulldozers had the effect of a herd of bison—disturbing the soil, removing the competing vegetation, but then going away. Unlike suburban lawns, detention basins are not regularly mowed, which allows the bur oaks to thrive. The bur oak seedlings are outcompeting the grasses and other herbaceous plants, and, in the absence of any further disturbance, they will form a new stand of venerable trees. This is a key advantage that the venerable tree species have over other trees: they have evolved to tolerate the competition of grasses and cane.

Figure 7.6. Bur oak seedlings growing in a suburban detention basin.

We also find the occasional venerable tree saplings along fencerows, though not in great numbers. These trees do not usually have a promising future. They are likely to grow into the fence wires and be wounded or girdled, and if they escape that fate, then they will be struck by mowers and trimmers.

One important characteristic of all the venerable tree species is that they don't reproduce well every year; rather, they reproduce at irregular intervals, a process called mast fruiting. The term *mast* is an Old English word for tree fruit, related to the word *meat*. Trees were said to produce soft mast, soft fruits easily eaten by animals, or hard mast, fruits protected by hard coats that required some work on the part of a person or animal to open. *Mast fruiting* simply means the intermittent production of large amounts of fruit. In our trees, a good mast year for one species is often accompanied by a good mast year for other species.

A heavy fruiting year is typically followed by one or more years with less fruit, sometimes no fruit at all. Mast fruiting has long been a mystery and therefore a subject of research for biologists. There are at least three reasons species may fruit heavily in some years and not at all in others.

Weather plays a fairly clear role in mast fruiting. A late-season frost after the trees are in flower may prevent pollen dispersal and fruit production. A spring drought may reduce fruit set, preventing fertilized seeds from maturing. A late summer storm may knock fruit off before it is mature.

The tree's internal reserves are also important in mast fruiting. Trees store large amounts of carbohydrate and protein as a reserve. A tree that is defoliated by an insect in the spring, for example, will immediately produce another set of replacement leaves using reserve nutrients. A very heavy fruiting year may deplete those reserves, leaving too little for the following season. This would produce a pattern of alternating good and bad years for fruiting, and this is consistent with what we see.

A third explanation for mast fruiting is that it serves a purpose in controlling the populations of squirrels and other seed eaters. Although the trees depend on squirrels and other small mammals to distribute seeds, very high populations of seed eaters may consume all the seeds. But if heavy seed crops are followed by one or more years of light seed crops, then the populations of seed consumers will remain low. It is a common observation that heavy seed years are followed by an explosion in squirrel populations, followed by a population collapse after a lean year.[2]

These explanations are not mutually exclusive, and all could be valid.

Regardless of the cause, it is quite clear that the venerable tree species have heavy and light seed years, and they sometimes produce no seeds at all.

Reproduction may occur only at very long, irregular intervals, which would require a good seed-producing year followed by a good year for growth. A heavy crop year followed by a year with a spring drought will produce few seedlings. Even under the best of conditions, few trees reproduce every year. For trees like bur oak and kingnut, many decades may pass between reproduction events.

The intervals between successful years of seedling production were made more complicated in our woodland pastures by the return of bison or other browsers to control the grasses. It may be that reproduction in the Bluegrass was quite a rare event.

Drought certainly plays an important role in reproduction. It is very common for trees to produce large seed crops following, or even during, a drought. It is also likely that grass and cane production would be reduced during or immediately following a drought. Bison might have left an area for several years during and after a drought. It is probably not a coincidence that the oldest trees in our area began growing right at the end of several prolonged periods of drought. We don't yet have enough data from tree ring analysis, but it is very likely that we will find a strong relationship between drought years and successful reproduction of trees in the Bluegrass and Nashville Basin.

Although these biological factors are important, reproduction today is extremely rare because of the poor conditions for tree establishment that we have created. Tree seedlings will not grow in mowed and herbicide-treated pastures, lawns, golf courses, and parks. They will not grow in asphalt or concrete.

Today there are far too few opportunities for trees to successfully produce seedlings to grow the next generation. As our existing venerable trees die from old age, development, and poor management, they are not being replaced.

Those few locations in which trees are successfully reproducing, at Griffith Woods, Julian Farm, Chandler Farm, Huskisson Farm, and Willman Way, provide important lessons for how to produce the next generation of trees. It is not that hard to create conditions in which our trees can reproduce. Intensive intermittent grazing at intervals of five to ten years is very successful. We could create reserves like Julian Farm where cattle are allowed to graze intensively for a season, then are removed for five to ten years to allow seedlings to grow. These reserves do not need to be large, as the tree pen at Chandler Farm shows.

To be successful in managing woodland pasture preserves for natural regeneration, we will need to learn to time intensive grazing activities to fit with the cycle of heavy seed years. There would be no point in intensively grazing a tree reserve after a poor seed year. We know that drought plays a role in the successful reproduction of our trees, but we don't know the details of timing, duration, and drought intensity that results in good reproduction.

There is a lot to be learned about managing woodland pastures to allow for natural regeneration. We will not be successful in assuring a future for our venerable trees until we understand more about how they reproduce.

8

The Guardians

Trees in Cemeteries

Venerable trees shade the Lexington Cemetery. I like to think of them as the guardians of the cemetery, serving the needs of the living as well as memorializing the dead. Among the thousands of trees in the cemetery are the largest known basswood (*Tilia americana*) and one of the largest known smoketrees (*Cotinus obovatus*). The basswood was here long before the cemetery was established, and nearby it are ancient bur oaks, chinkapin oaks, Shumard oaks, blue ash, and kingnuts that indicate that this was once a woodland pasture.[1]

Lexington Cemetery was established in 1849 and was the first cemetery in the region to embrace two new ideas—the perpetual care of both graves and land, and the use of a grove of old trees to create an attractive, bucolic landscape. Until the mid-1800s, people were buried either in churchyards or in family plots on private land. Burial grounds were established around the edges of large cities. Graveyards were often unkempt and unsanitary, and graves were not permanent. Little thought was given to making graveyards attractive, especially those for the poor. As cities grew, the space available for burials was not adequate to the demand, and graveyards were increasingly a public health hazard. City leaders began to seek more sustainable means of caring for the dead just as the dawn of the Victorian era (1820) was beginning to change our views of landscapes.[2]

The Victorian era ushered in a romantic vision of landscapes that began

Figure 8.1. The largest known basswood, next to the monument to Henry Clay, in Lexington Cemetery. The presence of this presettlement tree was noted at the time the cemetery began development.

Figure 8.2. A group of presettlement bur oaks on the edge of Lexington Cemetery.

replacing the more pragmatic views of earlier periods, and the style seized on as a template was the English manor-house garden. Even the idea of a garden was changing. Every early home in Lexington had a garden for subsistence, producing fruits and vegetables for its residents and, if there was excess, to sell. Soon, as the standard of living rose and homeowners could afford to buy food, the purpose of the garden began to change. Beginning in the 1820s, Lexington, along with other cities, became enamored of the landscaping movement, and its citizens began planting ornamental plants where once there had been crops.

Henry Clay played no small role in this transition. He began buying farmland on the eastern edge of Lexington in 1804 and established his home, Ashland, in the midst of a woodland pasture dominated by blue ash and white ash—the blue ash trees at Ashland today are remnants of the original woodland pasture. Clay was a progressive farmer, introducing new breeds of cattle and crops and becoming an early adopter of the new landscape movement. He graced Ashland with extensive plantings of trees, shrubs, and flowers, some collected by his gardeners in nearby woods, others imported from Europe. It is likely that he introduced many plants that later became popular throughout Lexington. Although he is widely credited with introducing ginkgo (*Ginkgo biloba*) trees, there is no evidence that he knew of this Asian tree, which probably was not available until late in the nineteenth century.

Many of the fine Bluegrass mansions and horse farms were established in the mid-1800s, soon after Henry Clay set up his estate. The landscape of these mansions was influenced by Clay's example and by the increasingly popular magazines featuring English gardens. The new profession of landscape architecture, pioneered by Andrew Jackson Downing and Frederick Law Olmsted, was beginning to influence the design of large spaces both public and private.

The landscape movement soon began to influence the design of cemeteries. As cities grew, churchyards and family plots were outgrown and larger cemeteries were needed.

Newer cemeteries, influenced by the landscape movement, were established in rural settings, or a rural atmosphere was created, to give a sense of sanctuary, rest, and beauty. This new vision of the cemetery was for the living as much as the dead. Cemeteries served both the function of a memorial to the dead and a place for the living to experience a taste of nature. On Sundays, cemeteries were crowded with families in their finest clothing strolling through the shaded paths.

Figure 8.3. An ancient Shumard oak growing around a headstone in Lexington Cemetery.

The first rural cemetery was the Mount Auburn Cemetery in Cambridge, Massachusetts. It was established not by the clergy but by the Massachusetts Horticultural Society. Mount Auburn was created in a seventy-two-acre mature hardwood forest, which formed the initial collection of trees, to be supplemented by horticultural collections of trees, shrubs, and flowers. Mount Auburn was dedicated on September 24, 1831, in front of an enthusiastic crowd of more than two thousand Bostonians. Mount Auburn was an instant success, and cities all over the United States began establishing their own rural cemeteries, using Mount Auburn as a model.

The establishment of these rural cemeteries, or cemeteries as natural areas, coincided with the rise of professional horticulture and landscape design. Cemeteries were seen as places to establish verdant landscapes covered not merely with grass like the old churchyard, but with flowers, shrubs, hedges, and, especially, trees. As cities matured, the idea of urban parks began to supplement the cemeteries, and by the end of the nineteenth century, cemeteries were once again primarily places for the dead. The exception to that is the small number of cemeteries that are so distinguished in their landscaping that they have remained a major draw for the public. Such is the case with the Lexington Cemetery, which most residents regard as the finest park in town.[3]

Several rural cemeteries were established in Kentucky and Tennessee in attempts to emulate the success of Mount Auburn. Few of them, however, had the advantage of being developed within an existing woodland pasture. Cave Hill Cemetery in Louisville was established on heavily worked farmland and a quarry. Over the years, trees have been added, but Cave Hill has never been able to emulate the Mount Auburn model fully. The Nashville City Cemetery is today a nationally recognized arboretum and is on the National Register of Historic Places. It was established in 1822 on farmland to the south of town. Similarly, the larger Mount Olivet Cemetery in Nashville was established on farmland. To my knowledge, the Lexington Cemetery is the only one in our region that imitated the Mount Auburn model by developing within existing woods.

Lexington Cemetery was designed to fit within the Victorian rural cemetery model. The developers of the cemetery were clearly aware of the rising landscape movement. Henry T. Duncan, one of the founders, was a regular reader of Andrew Jackson Downing's journal the *Horticulturist,* which frequently discussed rural cemeteries and landscape design. The cemetery's founders were able to purchase a forty-acre tract known as Boswell Woods that was very close to the heart of the city.

There are few descriptions of Boswell Woods before the development of the cemetery. It was said to be used for hunting and as a playground by local children. From the remaining trees, we can speculate that it was probably a woodland pasture. There are quite a few very old trees that are almost certainly remnants of Boswell Woods, including bur oak, chinkapin oak, Shumard oak, blue ash, and kingnut, as well as the enormous basswood. There are several very large sugar maple trees that could have been part of the original Boswell Woods.

The cemetery designers preserved as many of the old trees as possible. Roads were designed to wander among the trees. There were too many large trees to provide adequate room for graves, so some of the trees were removed.

The cemetery became a popular place for residents and visitors to wander and enjoy the cool shade and the monuments. At a time when there were no public parks to speak of, the cemetery served the needs of the citizens of Lexington, both living and dead.

The cemetery's bur oaks were the most prominent trees, and many

The St. Joe Oak.

A bur oak on a horse farm.

A remnant bur oak in a housing development.

A remnant bur oak in a commercial area.

Ancient trees on a farm in Fayette County.

An old bur oak on a farm in Bourbon County.

An old bur oak on a road in Fayette County.

A very old blue ash with horses in Fayette County.

A bur oak in a tree pen, Boyle County.

A blue ash in a field in Bourbon County.

A kingnut in the Hamburg Giant Grove, a suburban development in Lexington that is home to many large trees.

A leaning bur oak, Bourbon County.

A bur oak in the Hamburg Giant Grove.

A view of the Hamburg Giant Grove, Lexington.

Blue ash trees in Ecton Park, an urban park in Lexington. There are also blue ash and bur oak trees throughout the neighborhood, remnants of an old woodland pasture.

A blue ash with an old lightning strike, providing wildlife habitat.

The mother tree kingnut, Griffith Woods Wildlife Management Area.

A dead chinkapin oak, Griffith Woods Wildlife Management Area.

An old field with venerable trees, Griffith Woods Wildlife Management Area.

A chinkapin oak, Griffith Woods Wildlife Management Area.

Remnants of an old woodland pasture near Coldstream Farm.

Remnants of an old woodland pasture near Coldstream Farm.

Trees at Elmwood Stock Farm.

Trees at Elmwood Stock Farm.

Trees at Elmwood Stock Farm.

Trees at Elmwood Stock Farm.

Trees in the Loudon Grove.

Trees in the Loudon Grove.

Trees in fog, Fayette County.

Two views of a bur oak on the University of Kentucky campus.

A bur oak at Commonwealth Stadium, University of Kentucky. Decades of lawnmower damage have severely damaged the tree.

The Ingleside Oak.

Trees from the original Ingleside grove near the Ingleside Oak.

Intricate patterns in a dead blue ash. Dead trees provide important habitats for wildlife.

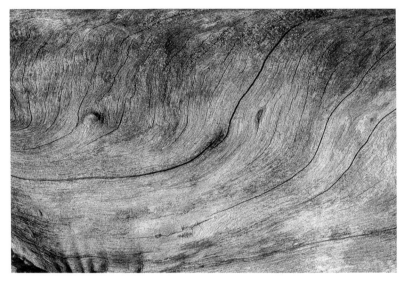

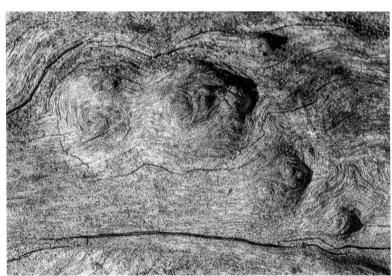

Bison, the original grazing animals of the Bluegrass, with a blue ash.

Horses, along with cattle, are the current grazing animals of the Bluegrass.

A leaning bur oak
at a Woodford County farm.

A dead blue ash at Griffin Gate,
part of the Coldstream complex.

Ancient trees in tree pens,
Bourbon County.

Bur oaks at a farm in Woodford County.

A bur oak in Woodford County.

A leaning bur oak, Woodford County. This tree is quite stable and has stood like this for many years.

Woodland pasture in the Oak Hill development, Woodford County.

Blue ash trees in a horse paddock, Woodford County. Horses rub the trees' bark, but the trees are not harmed.

Dead trees, a result of soil disturbance in a housing development.

A blue ash in Castlewood Park.

A chinkapin oak in an industrial area, Clark County.

A bur oak (*left*) and chinkapin oak, Boyle County.

Ancient trees in a pasture, Jessamine County.

Ancient trees in a pasture, Woodford County.

A bur oak, Woodford County. The tree lost its top, probably to lightning, and has built a new crown.

An old bur oak near Nashville.

The Llama Tree near Nashville.

The Old Schoolhouse Tree before a preservation plan was implemented.

The Old Schoolhouse Tree after clearing adjacent trees and shrubs as part of the preservation plan.

A bur oak adjacent to a highway, Fayette County. This tree is part of the Coldstream complex.

Bur oaks in tree pens, Woodford County.

Trees on a horse farm, Fayette County.

Veterans Oak at Veterans Park, Fayette County.

Blue ash trees, nearly dead, at Griffin Gate, part of the Coldstream complex.

Ancient trees in a pasture near a fence surrounding a sinkhole, Woodford County.

Cattle and ancient trees in a pasture, Woodford County.

A bur oak, Boyle County.

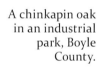

A chinkapin oak in an industrial park, Boyle County.

A chinkapin oak, Boyle County.

A kingnut in a housing development, Lexington.

A bur oak in Versailles, Woodford County, at the original toll road gatehouse.

Live and dead trees at Griffin Gate, part of the Coldstream complex.

A remnant bur oak in Fayette Mall, Lexington.

A remnant bur oak in the Coldstream development.

A remnant chinkapin oak in a commercial area, Lexington.

A remnant bur oak in a residential development, Fayette County.

An ancient bur oak, Lexington Cemetery.

A bur oak at Dixie Elementary School, Lexington.

Figure 8.4. Planted bur oaks in Lexington Cemetery. The cemetery has planted bur oaks since its founding in 1849, including this allée of trees that date from the mid-nineteenth century.

people requested that they be buried next to one of these trees. The cemetery managers recognized that bur oaks were the signature tree of the cemetery. Soon after the cemetery's incorporation, the horticultural staff began raising bur oak seeds from acorns collected on the property in their own nursery. Planting of bur oaks has continued ever since. Today there are bur oaks ranging from ancient presettlement trees to saplings planted last year. The cemetery now includes 107 large bur oaks greater than two feet in diameter, of which about 40 appear to be presettlement trees. Other bur oaks are quite old, but they appear to date from just after the establishment of the cemetery. Among these trees are several fine rows of old bur oaks along the cemetery roads.

There are fewer of the other venerable tree species, perhaps because the cemetery staff favored the bur oaks. There are several exceptionally large kingnuts, many blue ash, and a few chinkapin and Shumard oaks. There are also a number of red oaks of uncertain parentage, which is common throughout the region. The huge basswood is so large that it must have been here before settlement, but basswoods are not common in woodland pastures today.

The cemetery has planted Shumard oak throughout its history, although many of these came from nurseries and may not represent the local population of Shumard oak. Blue ash, chinkapin oak, and kingnuts have not been planted except for some very recently introduced blue ash saplings.

The cemetery's landscapers planted many other trees over the years, making it a very fine arboretum that represents native and imported ornamentals. The cemetery operates its own nursery, so the young trees being planted are the progeny of some of the venerable trees now in the cemetery. The cemetery also is a source of bur oak acorns for the Kentucky Division of Forestry state nursery.

Conditions in the cemetery are ideal for venerable trees. The cemetery managers have a long-standing policy of not using lawn chemicals, including fertilizer and herbicide. Soil compaction is carefully avoided. As new sections of the cemetery have been opened over the years, trees have been carefully preserved. The care with which the cemetery tends the land for the dead has favored the growth and longevity of the venerable trees.

9

The Loudon Grove

Trees in Public Spaces

By the late nineteenth century, cities were booming as the rural population moved to take advantage of the economic opportunities offered by urban living. As cities grew increasingly crowded, the demand for urban recreation and green space grew. Cemeteries were important, as we saw in the last chapter, but could not fully meet the demand for urban parks. Cities began looking for land to develop into parks, but because of rapid population and industrial growth, such land was hard to find. Cities in the north had begun setting aside parkland early in their development. Boston Common, the first public park in the United States, began as a shared cow pasture for all the residents. After more than two hundred years of conflict and overgrazing, Boston ended its use as pasture and gradually established it as a public park. Central Park in New York, the first major landscaped park in an American city, was created in the 1850s as the city of New York bought up prime land for a park. Just as Mount Auburn was the model for rural cemeteries throughout the country, Boston Common and Central Park served as models for public parks in the rest of the country.[1]

Southern cities were slower to recognize the importance of public parks. Finding land within the urban area for park development was a challenge. By the time Lexington began developing parks, the only urban properties available were old mansions and their grounds, the farmlands having been sold off for housing development.

Figure 9.1. Blue ash and other venerable trees surround the Loudon House in Castlewood Park.

One such property was Loudon House, a mansion built in 1850 for a cousin of Francis Scott Key, the author of the U.S. national anthem. Consisting of sixty acres in what was already suburban Lexington, the property was not one of the premier horse and cattle farms, but a bucolic retreat for its owners. Like Ingleside Manor and other stately homes, Loudon House was built in the middle of an existing woodland pasture.

By the early 1920s the house was no longer in use, and the city of Lexington acquired the home and about half of the original property for one of the first city parks. Loudon House remains today, a fine example of Gothic Revival architecture, now home to the Lexington Art League. The grounds surrounding the house became Castlewood Park, with a swimming pool, basketball and tennis courts, a baseball field, and a playground.

The trees that were there long before the mansion was built are there today. Some of the largest trees in the city surround the mansion and shade the ball fields. Bur oak and blue ash are the most numerous, and there are several fine chinkapin oaks.

Other parks in the older parts of the city were also established by the purchase of mansions and parts of estates. All of them still have some of the venerable trees that were there long before the estates were established.

Woodland Park, the oldest park in Lexington, has no mansion, but it was part of the estate of Henry Clay's son-in-law, James Erwin, and was known as Woodlands. In 1882 one hundred acres of the estate were used to build homes, and fifteen acres were kept for the private park. Woodland Park was purchased by the city in 1902. It is Lexington's most popular park and today includes a swimming pool, playground, baseball diamond, and skate park. In spite of its heavy use, the park has maintained a canopy of fine trees, including a number of rare horticultural specimens. There are also many bur oaks and several blue ash and chinkapin oaks. Some of these trees certainly date to presettlement times, but there are also quite a few large trees that must have been planted after settlement, though we don't know when.

As Lexington grew and demand for parks increased, the city began acquiring land outside the developed areas. These were mostly horse and cattle farms lacking the huge old mansions and the stately trees. Among the newer parks, only Masterson Station Park includes large, old trees. Masterson Station was one of the earliest estates in the Bluegrass, originally settled by Richard Masterson in about 1788 and sold to his brother in 1795. The estate remained intact until it was purchased by the federal government in about 1900. It subsequently became a treatment center called the Narcotic

Figure 9.2. Woodland Park includes ancient, presettlement trees that became part of James Erwin's Woodlands. Trees have been planted continually since the establishment of the park in 1882. These bur oaks are probably about eighty years old.

Farm, until parts of it were transferred back to the city in 1974. There are venerable trees around the edges of Masterson Station Park, especially on the north side, and on the adjacent property that was retained by the federal government.

Figure 9.3. An old bur oak at Masterson Station Park, a remnant of a former woodland pasture.

One of the most familiar public trees is the Veteran's Oak, growing near a popular walking trail and accompanied by a large interpretive sign in Veteran's Park. This exceptionally large tree is in the floodplain of West Hickman Creek, where it gets regular irrigation from floods. This is one of

Figure 9.4. This old bur oak is at Dixie Elementary School in Lexington. About twelve other old trees are in a nearby park.

the few large venerable trees in a dense forest. Surrounding trees are beginning to grow up into the canopy of the Veteran's Oak, and the forest should be thinned to keep from shading out its lower branches.

Parks in other Bluegrass and Nashville Basin towns and cities seem to have very few venerable trees left from old woodland pastures. Smaller cities did not establish parks where there were existing remnants of woodland pastures or old mansions. Some of the most important historic sites in the Bluegrass, such as Fort Harrod in Harrodsburg, Constitution Square in Danville, and Perryville Battlefield, are devoid of presettlement trees. Fort Harrod is home to an exceptionally large Osage-orange, but that is a planted tree.

Certainly parks in many cities include large, old trees, but most of these were planted during the creation of the parks. Lexington is unique among cities in having parks that were built on the sites of mansions that in turn were built on the sites of ancient woodland pastures.

There are venerable trees at a few schools. Dixie Elementary School

Figure 9.5. An old chinkapin oak sits along a creek at Lipscomb Academy Elementary School in Nashville. Several other ancient trees are in woods along the edges of the school yard. Parks and yards around this property contain dozens of ancient trees.

was far outside the urban area of Lexington when it was founded in 1966. Today the area around the school has been fully developed, but Dixie sits on a large campus that is both a school yard and a park. The school yard includes a number of old trees, and there is also a row of trees preserved along a former property line. Right next to the school is a magnificent old bur oak. Someone had the foresight to protect this majestic specimen tree during construction of the school.

At Lipscomb Academy Elementary School, in the southern suburbs of Nashville, a stately old chinkapin oak stands along a creek. The neighborhood around this tree includes a few ancient bur oaks, blue ash, chinkapin oaks, and a kingnut, indicating that this tree was part of a woodland pasture.

Venerable tree species are unable to regenerate naturally, at least not in adequate numbers. If we want to have these trees in our future landscape, we will need to plant them. Much of the region is now urban or will become urban in the coming years. Where will we find the space to plant and maintain venerable trees to continue the legacy we were given by early settlers?

All the trees we are discussing require a lot of space to grow and thrive,

Figure 9.6. An old bur oak street tree.

Figure 9.7. The bur oak shown in Figure 9.6 has inadequate root space, even after the sidewalk was moved.

and for this reason they are not suitable as street trees. A few venerable trees have been left behind and become street trees, but they rarely do well. Tightly constrained between street and sidewalk, in heavily compacted soil, they eke out an existence. Bur oaks, among the toughest trees in our landscape, may be able to survive, but the street is not a good habitat for a large tree.

Our trees may be tough, able to survive droughts that kill most trees, but they are very sensitive to soil compaction and excessive fertilizer and herbicide. Development of suburbs often causes the decline of venerable trees that are left behind because of compaction, grading of the landscape, and the heavy use of chemicals typical of the suburbs.

Large, open landscapes such as school campuses and parks are good candidates not only for preserving existing old trees but also for planting their replacements. Soil compaction from recreational use is a potential problem, but it can be mitigated with good soil management, especially if we avoid the use of heavy lawn mowers. We need to manage these campuses and parks with minimal or no use of fertilizer and herbicide. Organic lawn care methods are designed to maintain soil health and are beneficial to trees. Given a choice between managing for luxuriant grass and ensuring the survival of ancient trees, we should always favor the trees.

School campuses and parks also provide a great opportunity to engage the public, especially children, in nurturing existing venerable trees, planting replacement trees, and perhaps even restoring the original woodland pasture habitat. Yet most of our school yards are devoid of trees, or they contain sad little trees poorly planted and poorly maintained. How much better it would be to engage students, parents, and teachers in planting trees that have cultural significance and caring enough about them to ensure their future. These trees are our heritage, and their existence helps connect us to our past.

We need to find other large landscapes in urban areas where venerable trees already exist and can be nurtured or where seedlings of the same species can be planted. Landscapes that consist of large, open space but are of low recreational use include the campuses of houses of worship, corporations, and colleges.

Many houses of worship are set on large, landscaped campuses. Some of these have individual venerable trees or groves of presettlement trees. The national champion Ohio buckeye is on the campus of the Catholic Diocese of Lexington. Churches in many smaller towns in the Bluegrass and

Figure 9.8. An old blue ash in front of a church in Lexington.

Figure 9.9. A church with more than two acres of grass in Danville, Kentucky. This landscape provides great opportunities for creating new groves of venerable tree species.

Nashville Basin have old presettlement trees on their properties. The Wilmore Camp Meeting Grounds in Jessamine County is a place of worship established in the middle of a grove of venerable trees.

 Church, synagogue, and mosque properties also provide opportunities for planting venerable tree species and even for restoring entire woodland pastures. With large, underused properties and congregants who can be enlisted for volunteer efforts, it should be possible to make houses of worship into sanctuaries not only for people but for venerable trees. The creation care movement in Protestant churches seeks to engage congregations in environmental improvement. Religious organizations and their congregants could play a key role in restoration of our environmental heritage, creating new woodland pastures of native trees to replace those that are lost.

 Corporate and college campuses manage thousands of acres of lightly used land that could be put to use to establish large woodland pastures. Today many corporate campuses are sterile landscapes of grass. College administrations do a pretty good job of planting and maintaining trees on

Figure 9.10. The Kissing Tree at Transylvania University in Lexington.

their lands. It is part of the charm of a college or university campus to have academic buildings surrounded by large shade trees. Some of these trees are significant components of campus life. The Kissing Tree at Transylvania University is a huge white ash with a wooden bench around its trunk. It is not a presettlement tree but probably became established either naturally or by planting some time in the 1800s. In the years before 1960, when public displays of affection were not tolerated on college campuses, the Kissing Tree was the one place where holding hands and discreet kissing were not met with a rebuke. Although its role as a facilitator of romance is not as critical today, the Kissing Tree is revered as part of the rich history of the college.

Among college and university campuses in the Bluegrass and Inner Nashville Basin, only the University of Kentucky campus has a large number of venerable trees. Much of the campus was probably a woodland pasture before the development of the university. Centre College, in Danville, has quite a few large, old trees but only one bur oak that could be a lone representative of a woodland pasture. Other campuses in the region have

Figure 9.11. A prominent ancient bur oak on the University of Kentucky campus.

large trees, but they show no evidence of including remnants of the original woodland pastures.

The University of Kentucky has continued to plant bur oaks throughout its history, so that there is a wide range in the size and age of bur oaks. A walk around the oldest, central part of campus provides clear evidence of its woodland pasture history, as there are several huge bur oaks and some blue ash and chinkapin oaks. Many of the oldest trees have disappeared in recent years, some to construction and some to poor management. Most of the oldest trees on campus are suffering from chronic mower injury and other stresses, and they will probably not survive much longer. Some of these trees present a hazard and need to be removed.

A very unfortunate example of neglect of a magnificent grove of trees is the campus of Sullivan University, a private college with a campus in Lexington on Harrodsburg Road. The site was once the home of Hal Price Headley, a renowned horseman and owner of Beaumont Farm, which once covered more than four thousand acres that is now largely developed. There are scattered remnants of the original woodland pasture throughout the

Figure 9.12. One of two bur oaks next to Commonwealth Stadium. Many tailgate parties have taken place under this tree.

Figure 9.13. The base of the tree shown in Figure 9.12 has sustained extensive mower damage and decay. The fungus is *Laetiporus cincinnatus,* sometimes called chicken of the woods. The fungus causes decay of heartwood. By itself, the fungus would not be fatal, but ongoing, repeated mower injury has severely wounded the tree. Note the hole caused by boring beetles. Other wood decay fungi are present as well.

Figure 9.14 A declining grove of venerable trees at Sullivan University. This grove once surrounded the home of the renowned horseman Hal Price Headley.

property. The St. Joe Oak was once part of the Beaumont woodland pasture. Headley built his home in a dense grove of venerable trees and took excellent care of them. Eventually, the farm was developed. Headley's original house was removed to make room for a modern office building that was used first by a coal company and later by an engineering firm. Both the coal company and the engineering firm took excellent care of the trees. I visited the campus often, admiring the exceptionally large trees, which included all the venerable tree species. There were also younger kingnuts, white ash, and other native trees.

In recent years, Sullivan University has been managing the property, and the trees are rapidly failing. The cause of this failure is an intense effort to grow dark green, weed-free grass. The use of large amounts of fertilizer and pesticides necessary to maintain a pristine lawn in such a shady environment is very stressful for old trees, and many are in decline. The school also failed to treat its very large white ash trees for emerald ash borer, and they all died in 2013. It is not too late to rescue the remaining trees, but it will require the school to put emphasis on the trees and not on the grass by shifting to an organic lawn care strategy that is tolerant of clover and other "weeds" and minimizes the use of lawn chemicals.

Figure 9.15. A chinkapin oak above Clark's Run at the RR Donnelley campus in Boyle County, Kentucky.

The corporate campus of RR Donnelley, a printing plant in Danville, Kentucky, is an excellent example of both the maintenance of venerable trees and the potential for the creation of new woodland pastures. The campus of this printing company is located in an industrial park south of Danville, on the southern edge of the Inner Bluegrass. To the west of this campus there are extensive groves of venerable trees on farms in Boyle County. Clark's Run, a lovely little creek, meanders along the edge of the property, and there are several large, presettlement trees along the creek. Uphill, toward the printing plant, are several more, including chinkapin oaks and blue ash. Most of the campus is grass. Like the grounds of many corporations throughout the region, the extensive mowed lawn could provide an excellent opportunity for planting groves of venerable tree species and restoring the woodland pasture habitat. Such a restoration would beautify the campus, provide shade, and allow the long-term conservation of our heritage of woodland pastures.

Campuses, whether religious, academic, or corporate, provide an opportunity to plant saplings of native Bluegrass trees instead of the ornamentals currently in use. Some of these campuses are large enough that they could, with proper selection of planting material, come to resemble the

Figure 9.16. Newly planted trees on the University of Kentucky campus, including bur oak and chinkapin oak.

original woodland pasture landscape. Like most landscapes, however, these campuses are becoming homogenized with the products of modern industrial horticulture; there is too much emphasis on grass, and the genetic base of trees has narrowed. It is cheaper to plant red maples or Texas Shumard oaks grown from cuttings than seedlings of native trees. But these modern plantings do not last long—many of them last only a few years and have to be replaced. Managers of large campuses should look at the long-term costs and benefits of tree planting and maintenance. Slow-growing, long-lived trees may not provide instant shade, but the long-term benefits of large trees and the low cost of maintenance make them a bargain. A recent planting of native trees including bur oak and chinkapin oak at the University of Kentucky is a promising start to the use of campuses for re-creating the woodland pasture habitat.

10

The Coldstream Tree

Groves, Remnants, and Developments

The Legacy Trail is a twelve-mile bike and walking path that winds along Cane Run from downtown Lexington to the Kentucky Horse Park. Close to the trail is an old bur oak, its top dead but growing vigorously next to a stand of cane. The tree is named for Coldstream Farm, a livestock farm that later became a university research farm and today is a research park operated by the University of Kentucky.

Like most Bluegrass farms, Coldstream Farm has passed through many identities and owners. Coldstream Farm was formed in 1915 from part of McGrathiana Farm, a renowned horse and cattle operation that itself was built from several farms in 1873, and those farms dated back to the earliest days of settlement.

It is a common experience in searching for venerable trees to come upon a small cluster or a single isolated tree, like the Coldstream Tree. At first visit, these isolated trees seem to have been always alone, not part of a larger grove, but that is almost never the case. Instead, the isolated trees represent remnants of a woodland pasture. Often, when I find a single ancient tree, a search of the countryside or neighborhood around the tree will reveal more.

I knew that the Coldstream Tree had not grown there alone because I had seen many other trees when I first visited Coldstream Farm in 1982. At that time, Coldstream Farm was a university research farm. There were

Figure 10.1. The Coldstream Tree along the Legacy Trail. There is another venerable tree on the other side of the creek, and a stand of cane to the left.

Figure 10.2. Venerable trees in a field near the Coldstream Tree. This farm is prime development land within the Urban Services Boundary of the city. There are at least fifteen venerable trees around the edges of these fields.

dozens of blue ash; several chinkapin oaks, bur oaks, and Shumard oaks; and a few kingnuts. All were in poor condition from years of soil compaction and mechanical injury by tractors used to mow the pastures. The farm managers and the dean of the College of Agriculture were unwilling to make changes to management practices that would extend the lives of these trees. Over the next thirty years, all but a few of these trees died.

There are many venerable trees on nearby properties. As I began to explore the land near the Coldstream Tree, walking through corn stubble and around apartment complexes, I found more trees, each of them set apart from the others, separated by plowed fields or apartment buildings. I counted one tree, then another, until soon I had found dozens.

After a bit more exploration, I realized that these trees are the remnants of a very large woodland pasture that must have extended over several square miles, far beyond the bounds of Coldstream Farm, and I have since located and mapped many trees remaining from the original woodland pasture.

The largest group of remaining trees in the Coldstream grove stands in front of a hotel on the main road, another old buffalo trace. The grove surrounding the Griffin Gate Resort development in Lexington is the finest remaining grove of ancient trees inside the urban boundary of Lexington. The Griffin Gate trees are on a lawn surrounding a hotel and residential complex; an 1854 mansion still stands behind them. Dozens of blue ash, along with several bur oaks, chinkapin oaks, Shumard oaks, and kingnuts, are scattered across the lawn leading down to the main road. Some of the trees are dead or near death and have been left standing: in falling, they would do no harm. The lawn gets little human traffic, so compaction is not a problem. This grove should last a long time, although as the trees die, they are not being replaced with the same species. High-intensity lawn maintenance, including the use of fertilizer and herbicides, threatens the long-term health of these trees.

The rest of the trees in the Coldstream area are scattered and increasingly solitary. As development proceeds, these trees may disappear one by one, until only the Coldstream Tree remains. It is the only one of these trees on public parkland along the Legacy Trail and therefore protected from development.

The more we learn about the location of venerable trees, the more we can see that none of them began as solitary trees in a meadow or along a stream. They were all components of woodland pastures that stretched over

Figure 10.3. A grove of venerable trees at Griffin Gate Resort, near Coldstream.

large areas. When we find solitary trees or small groups of trees in urban settings, we can be sure that they are remnants of woodland pastures that were left on their own as development took away their neighbors.

As is the case with the Coldstream Tree, we can sometimes find evidence of the original grove and its boundaries by exploring the land or by referring to old land ownership records. The Ingleside Oak stands alone on Harrodsburg Road, but Thomas Knight's photographs from 1904 show a vast grove around Ingleside Manor. With a bit of exploration, I found several other trees that were part of the Ingleside grove still standing. One tree is at the back of a vacant lot, another behind a fraternity house, several more in the front and back yards of houses built long after Ingleside Manor was gone.

There are many old trees in housing developments in urban and rural areas throughout the Bluegrass and the Nashville Basin. In a few neighborhoods, groups of trees have been carefully preserved. This is easy to do when there is a creek or detention basin that cannot be developed. A neighborhood in Lexington called Squire Oak has a large group of very old trees

Figure 10.4. A bur oak in a parking lot near the Ingleside Oak.

Figure 10.5. A bur oak in a front yard near the Ingleside Oak.

Figure 10.6. A bur oak in an abandoned lot near the Ingleside Oak. This tree and the ones shown in Figures 10.4 and 10.5 were all part of the original woodland pasture that became Ingleside Manor (see figures 5.1, 5.2).

Figure 10.7. The Blackford Oak.

Figure 10.8. Dead trees near the Blackford Oak.

Figure 10.9. A very old bur oak in a yard in Harrodsburg, Kentucky. There are several similar trees in the same neighborhood.

mixed with younger trees along a creek. This grove has become a centerpiece of the neighborhood, and homeowners are taking good care of the trees.

In another neighborhood called Blackford Oaks, a single tree was preserved at the request of the original landowner. It has not been maintained properly, however, and is in decline. Other old trees in the same housing development have been seriously damaged by grading and are dead or dying. It is not sufficient merely to leave a few trees behind during development—they need to be cared for.

Smaller towns and cities in the Bluegrass and Nashville Basin include a substantial number of individual ancient trees in older neighborhoods. In cities such as Harrodsburg, exploration of these neighborhoods turns up several dozen old trees, often in adjacent yards. The lower intensity of development and large yards result in trees being able live out their natural life spans without the soil compaction and mower damage so common in larger urban areas. The lightly developed southern suburbs of Nashville are also a treasure trove of isolated old trees. The ancient trees, mixed in with more recent plantings in large suburban yards, are probably remnants of once-extensive woodland pastures.

Figure 10.10. A magnificent Shumard oak shades a suburban yard in Brentwood, Tennessee.

Figure 10.11. Part of the Hamburg Giant Grove, an extensive area of scattered, very large trees in a residential and commercial development. These trees remain from a woodland pasture that occupied several square miles before development.

Figures 10.12. Oak Hill development, Woodford County, Kentucky.

Figures 10.13, 10.14. Oak Hill development, Woodford County, Kentucky.

There are some wonderful groves remaining even in developed parts of cities and towns in the Bluegrass. The Hamburg development, a large horse farm developed into housing and shopping areas in Lexington, has preserved many venerable trees in or near floodplains, where they will be undisturbed. The Hamburg area is only partly developed, and there are many trees remaining in the existing farmland that could be preserved and become landmarks for their neighborhoods. I have come to refer to this area as the Hamburg Giant Grove because it includes dozens of exceptionally large trees, some of the largest remaining trees in Fayette County.

The best example of an intact woodland pasture in a developed area is in Woodford County, Kentucky. Oak Hill was a farm established in 1796. It has recently been developed under Woodford County's strict rural zoning, in which substantial areas of undeveloped land accompany houses on small lots. At Oak Hill there is an entire woodland pasture surrounding the houses. The property developers and home owners have been planting bur oaks to ensure the future of their remarkable development.

Old woodland pasture trees left standing in a development may not have an easy life. The main cause of decline and death of venerable trees that are left in suburban landscapes is the management of the suburban lawn. Home owners and managers of large properties often seem to have an obsession with bright green, weed-free grass. This kind of lawn bears no resemblance to the mixed native plants in which the trees evolved. Heavy doses of fertilizer, herbicide, and water are required to maintain lawns. Fertilizer and water stimulate growth in trees, and you might think that is good, but it isn't. Fast-growing trees have shorter lives than slow-growing trees. A tree that has grown for centuries without added fertilizer and with frequent water stress may not respond well to the new lawn-care regime. Herbicides can create chronic stress. As lawn care is increasingly taken over by landscape contractors, the heavy lawn tractors they use compact the soil and mechanically damage roots.

One compelling task is to find the remaining trees on land that is slated for development. Within the urban boundary of Lexington are several thousand acres yet to be developed, and much of this land is currently or recently farmland that includes a substantial number of venerable trees.

11

The Elmwood Trees

Growing Old Gracefully

As we get farther away from the urban centers, the groves become larger. Some are small groups of trees on a single farm. Many of them cover multiple farms, interrupted by roads or pastures from which trees have been cleared. These groves form a mosaic that, while less dense than the presettlement groves, maintain some of the features of the presettlement woodland pastures.

Most operating farms in the region focus on livestock, either cattle or horses. The horse farms attract the most attention, but cattle farms have always been an integral part of the farming community and were actually established before horse farming became popular. Cattle can have positive or negative effects on trees. Cattle confined to a single pasture can compact the soil, expose roots, and create erosion, but good cattle husbandry can also improve soil.

Long before Kentucky was known for its horses, the Bluegrass was cattle and sheep country. From the time of settlement until the late 1940s, farms in the Bluegrass and Nashville Basin were major sheep producers. There were over a million sheep in Kentucky in 1942, nearly all of them in the Bluegrass. Like cattle and bison, sheep are grazers rather than browsers and would not have damaged trees in woodland pastures. After World War II, sheep farming in the East declined because of the replacement of wool by synthetic fibers and the difficulty of managing a major sheep disease, foot

Figure 11.1. A woodland pasture at Elmwood Stock Farm. This is one part of a woodland pasture that covers most of Elmwood Stock Farm and several surrounding farms.

rot. Today sheep are making a comeback in Kentucky and Tennessee, and it is increasingly common to see flocks of sheep on our farms.

ELMWOOD STOCK FARM

Sheep, cattle, horses, and bison are well suited to woodland pasture habitats. As the old photographs show (see figure 5.2), most Bluegrass farms had pastures shaded by groves of venerable trees. Today the vast majority of pastureland lacks the shade that was so important to early farms. But at Elmwood Stock Farm in Scott County, cattle and sheep graze in the shade of woodland pastures as they have for hundreds of years.

Elmwood Stock Farm is a multigenerational family farm in Scott County that traces its roots back six generations. The Bell family today produces cattle, sheep, and a wide variety of crops. The farm is one of the largest certified organic farms in the region.

As organic farmers, members of the Bell family regard their main mission as good soil management, and their cattle, sheep, and crops as the product of the good soil. Mac Stone, a member of the Bell family, likes to say that Elmwood Farm numbers its livestock in the trillions, referring to the soil animals and microorganisms that maintain the fertility of the land.

The practice of rotational intensive grazing, in which cattle or sheep graze intensely in a small area until they eat down to stubble and are then moved to another pasture is good for the health of the livestock and the pasture. Along with good cover-crop management, rotational grazing creates ideal soil conditions for livestock, crops, and venerable trees.

Our soils are prone to compaction, whether from heavy mowing machines or overuse by livestock. Many former woodland pastures that are heavily grazed or mowed with large tractors have lost most or all of their trees because of soil compaction. In some of these pastures it is difficult to get a shovel in the soil without great effort, and you can imagine that it is not easy for roots to grow through this dense material. Rain tends to pool on the surface of compacted soils and run off quickly. In contrast, the soil at Elmwood Stock Farm is light and fluffy, easy to turn with a shovel. The soil has a rich, pleasant smell because of its high organic matter content and diverse community of soil organisms.

Not surprisingly, the trees in the woodland pastures at Elmwood Stock Farm are thriving. These trees are old, and they show it. Many have been hit by lightning and have lost their tops. Many show signs of decay. About

Figure 11.2. An old bur oak at Elmwood Stock Farm.

Figure 11.3. Many venerable trees have been struck by lightning. The dead spar at the top of this blue ash is characteristic of lightning-struck trees.

half of the blue ash trees have woodchuck (*Marmota monax*) holes in their bases. It is a common belief that woodchucks, also called groundhogs, make these holes by chewing the wood, but that is not true; in spite of their name, woodchucks don't chuck wood. The name derives from a Cree or Algonquian word for this animal, *otchek*. Woodchucks simply move into holes created by decay. Old trees are often wounded around their bases, sometimes by tractors or cattle, but by natural causes as well. As decay fungi move into the wound and consume the wood, holes open at ground level or below, providing habitat for woodchucks and other animals.

These trees also provide important habitats aboveground. The dead branches and hollow stems provide resting, feeding, and nesting sites for birds, including hawks and owls, and for small mammals, including squirrels and flying squirrels. Supporting these animals is a diverse community of insects and fungi that feed on deadwood without harming the living parts of the tree. We can think of each of these ancient trees as diverse communities of organisms supported by deadwood.

Once the entire tree is dead, it continues to provide important habitat, supporting a diverse range of organisms, including birds, mammals, beetles, fungi, bacteria, and others.

At Elmwood Stock Farm, dead trees are allowed to remain standing, and they are left when they fall if they are not in the way. The rich soil, intermittent grazing, ancient live trees, dead trees both standing and down probably create the richest biodiversity found in any of our woodland pastures.

Although the ground cover at Elmwood Stock Farm is not the same as the original native cover, the groves of trees there probably come the closest to the original woodland pastures that were here in 1779.

The famous horse farms of the Bluegrass, along with the less well-known but splendid farms of the Nashville Basin, provide excellent habitat for venerable trees. The low density of horses per acre of pasture helps minimize soil compaction. Horses do like to chew on bark, although they rarely damage the living inner bark. Most well-managed horse farms use fences to protect trees from the horses.

Calumet Farm is close to the Bluegrass Airport and Keeneland Race Track, along a major thoroughfare that was once a buffalo trace. It is probably the farm seen by the largest number of people. Unlike many of the farms that have long used black fences, Calumet Farm continues to use elegant white fencing. The combination of gorgeous horses, white fences, fancy barns, and gigantic trees on the rolling slopes of Calumet is hard to forget.

Figure 11.4. A standing dead blue ash. Dead trees that are not too close to buildings and roads are allowed to remain standing at Elmwood Stock Farm. They provide important wildlife habitats and are quite beautiful. This one is clothed in native Virginia creeper.

Figure 11.5. A dead tree in a pasture. Dead trees continue to serve a role at Elmwood Stock Farm. They provide wildlife habitats, and their decay contributes to soil improvement.

Woodford County, as we have discussed, is the epicenter of venerable tree groves in woodland pastures. The standard of care for trees on the county's farms varies, but the generally low-intensity management of pastures, combined with minimal soil compaction, makes it likely that the venerable trees will remain for a long time.

As long as the essentially rural nature of the Inner Bluegrass and Inner Nashville Basin is maintained, there are good prospects for large numbers of venerable trees to survive for their natural lifetimes. We don't know how long those lifetimes might be, but the vast majority of these trees are growing well and staying ahead of decay and decline.

When the trees die, they either are not being replaced or are replaced with other species. It is extremely unlikely that the replacement trees will live as long as the original ones—they are typically fast-growing, short-lived species and ornamental cultivars that have not stood the test of time.

The venerable tree species cannot regenerate naturally, even at Elmwood Stock Farm. The big difference between Elmwood Stock Farm today and before 1779 is this: although cattle are moved from pasture to pasture frequently, they return to each pasture within a few months, whereas the bison left for years at a time. It is quite clear that long intervals of five to twenty years are required for venerable trees to regenerate naturally.

If we are to have groves of venerable trees in our landscape for hundreds of years more, we need to have some means of orderly replacement with the same species, either from natural regeneration or from tree planting. The only way for natural regeneration to be successful is to exclude livestock from a section of land for long enough for natural regeneration to take place. For this purpose, tree pens could be ideal.

We will have the most success in maintaining the woodland pastures by planting trees. For this purpose, we will need to establish nurseries to grow seedlings collected locally.

Lightning and Trees

Watching a tree being struck by lightning is terrifying or beautiful, depending on how far away the observer is standing. I have seen many trees being struck by lightning while I was out in the woods working or hiking. One day, while waiting out a storm on a lake in the Adirondack Mountains, I saw a huge white pine explode into splinters. There was not a piece of wood left that was longer than a baseball bat.

Later, while living in Malaysia, where intense lightning is an everyday occurrence, I saw many trees being struck without apparent damage. One day, I was very close to a huge red meranti tree (*Shorea macroptera*) that was struck with blinding light and earsplitting noise, but the tree appeared unharmed. What were the differences between the white pine that was destroyed and the red meranti that was unharmed? Over the years, I have kept notes about lightning strikes that I have observed and the marks of lightning on trees.[1]

At Elmwood Stock Farm, most of the large trees have lightning scars. These range from raised ridges on the bark to deep cavities. Trees all over the property have been struck. Some of them are tall trees at the top of a hill, while others are shorter trees in more protected locations. Taller trees, trees higher on a slope, or trees close to a good conductor such as an electrical wire are more likely to be struck, but in an intense storm, most trees are at risk.

Electricity seeks the path of least resistance, and this explains most of the differences we see among trees. The red meranti tree that I saw being struck had smooth bark. Rain had already thoroughly wetted the stem before the strike, and it is likely that lightning traveled down the water layer

Figure 11.6. An old bur oak many years after a lightning strike. Cambial growth has closed the bark over the original injury.

on the outer surface of the tree. It is known that smooth-barked trees are less likely to be damaged by lightning than are rough-barked trees. The deeply furrowed bark of the white pine would take longer to become thoroughly wet, so it is more likely that the current would travel down the tree internally.

Within a tree, there are tissues that differ in conductivity. In the spring and early summer, the tree is actively producing phloem and xylem, and the cambial cells between the bark and the wood are moist and high in salt, making the cambium and its adjacent tissues excellent conductors. If lightning strikes a tree during active growth, then the path of current is most likely to be along the cambium, and this can have two results. Typically, current travels along a strip of cambium, blasting off a strip of bark and perhaps cracking wood underneath. The strip often curves down the trunk. This kind of strike is fairly benign. Cells adjacent to the dead cells will create a compartment and eventually form a ridge of bark.

Occasionally, current will travel down the entire cambial layer, instead of following a narrow path. In this case, the entire cambium will die, and this is fatal to the tree. I have seen a few trees in which all the bark popped off, but the wood was undamaged. In other cases, the cambium may be killed without any apparent outside injury, but the leaves will soon wilt and the tree will die.

Lightning strikes are more common in the Bluegrass and Nashville Basin in the late summer than in the spring. Once diameter growth has slowed down by late June, the path of least resistance is no longer the cambium but the wood. As current travels through the wood, water is instantly converted to steam, and it is the steam pressure and heat that do the damage. Lightning following the wood can crack the tree, pop some of the bark off, or cause the whole tree to explode. Explosions of the wood are more common in conifer trees, like the white pine I observed, because the high resin content of the wood provides resistance to the flow of current. I have seen only two trees in the Bluegrass explode on being struck, and both strikes occurred in late summer.

A tree with a substantial amount of wood decay is at greater risk for explosive damage than a sound tree. Decay columns in the stem are usually wetter than the surrounding wood and provide a path for current to travel, but decay columns are complicated, as they contain voids and pockets that vary in moisture content, which allow flashing of current between gaps and explosive development of steam.

Figure 11.7. A bur oak many years after a lightning strike that blew out the top of the tree. The tree grew a new crown. The main stem has been injured but continues to grow well, and there is little decay in the exposed wood.

Though lightning strikes can be quite dramatic, as wood and bark fly through the air, the most insidious damage may be to root systems. A tree struck by lightning that conducts electricity to the root system without any apparent damage aboveground may have suffered a fatal injury. Such a tree may stand quite robustly through the summer, or the leaves may wilt, but the following spring the tree will be dead. Because there is no external sign of injury, the death remains unexplained.

When a tree is struck by lightning, it is often tempting to treat the tree immediately by pruning off dead branches and bark or taking other aggressive actions. This is a mistake. If a tree is likely to injure someone or damage

Figure 11.8. A blue ash with an old lightning injury. Decay has opened up cracks in the stem, but the tree is growing vigorously. The cracks provide good wildlife habitats.

Figure 11.9. Sudden death in a blue ash. This tree failed to leaf out after growing normally the previous year. There were no external signs of damage or disease. The most likely cause is root death due to a lightning strike.

property, dead branches or the whole tree needs to be removed. Otherwise, it is best to wait at least a year before taking action. Dead branches may not be entirely apparent at first, and it is possible for the entire tree to die.

For trees in woodland pastures, the best action is no action. A lightning-struck tree that survives will develop decay columns that provide habitat for wildlife. Trees that lose their entire crown to lightning can rebuild, becoming a shorter but still vigorous tree.

12

The Floracliff Trees

The Long Lives of Venerable Trees

The oldest tree in Kentucky is a modest individual, not huge or dramatic. A chinkapin oak, it grows on a steep limestone bluff above the Kentucky River at Floracliff State Nature Preserve. When Neil Pederson and Beverly James first explored Floracliff for old trees, they thought little of the prospects of this tree. Pederson, a dendrochronologist at Harvard Forest in Petersham, Massachusetts, and James, the manager of Floracliff, were curious about several trees that were not large but looked old.

What does it mean for a tree to look old? We certainly cannot judge a tree's age by looking at it. On a rich, moist site, a tree may grow ten times as fast as the same species on a dry site. Very large trees are often much younger than small trees of the same species. Two trees that are similar in size may be vastly different in age.

Pederson developed a list of indicators of great age in trees. They include a cylindrical, rather than tapered, stem; a sinuous, rather than straight, stem; a small number of large-diameter, twisting limbs; a low crown volume; and a low ratio of leaf area to trunk volume. Pederson also said that the bark of very old trees was smooth or balding compared to that of younger trees, but I don't think this applies to two of our trees, bur oak and blue ash. To his criteria we add a tendency for leaves to be tufted at the end of branches. While these sound complicated, with a little practice most people can recognize a very old tree, or at least a candidate tree. Notice that

Figure 12.1. The oldest known tree in Kentucky, Woody C. Guthtree, 404 years old in 2015.

his criteria say nothing about size. Quite often the oldest trees in a stand are not among the largest.

Because trees leave a permanent record of their lives, the annual rings, it is possible to determine the exact, or approximate, age of a tree. To obtain those rings, you can either cut the tree down, which is often not desirable, or take a sample of wood using a thin metal tube called an increment borer.

To the surprise of Pederson and James, this tree turned out to be more than four hundred years old. It germinated in 1611, the year in which Shakespeare's *Tempest* was first performed and just four years after the founding of Jamestown, the first English colony in the New World. The year 1611 was a good growing season that came after several years of moderate to severe drought. Pederson named the tree Woody C. Guthtree. There are other very old trees in the same stand, several over three hundred years old. Very close to the oldest tree is a slightly larger chinkapin oak that is a mere 157 years of age. It is growing close to a creek and grew faster because of its greater access to water.

This does not mean that Woody C. Guthtree is older than any other tree in the Bluegrass, only that it is the oldest tree for which a highly accurate age has been determined. Other old trees in the Bluegrass for which age has been estimated include blue ash (405 years), bur oak (440 years), Shumard oak (480 years), American elm (410 years), and kingnut (200 years). The more accurate count by McEwan and McCarthy of blue ash showed a maximum age of 249 years. Accurate counts of annual rings exist for so few trees in the Bluegrass that we are likely to find much older trees as we obtain more samples. We have no published annual ring counts for trees in the Nashville Basin.[1]

Why are these trees able to reach such great age, and how much longer can they live? We don't really know the answer to either of these questions. The oldest trees in the world are over five thousand years of age, and the oldest hardwood trees in eastern North America are over six hundred years. Oaks in England are known to have lived more than one thousand years.

It is possible that there are trees in our landscape that are over one thousand years of age and we just haven't found them. We may never find them. The vast majority of very old trees are hollow and decayed. The oldest rings have been lost and cannot be recovered. All we can say for certain is that we live among very old trees.

We constantly see accounts in the news media of "a three-hundred-year-old bur oak" or we hear self-styled experts extol the virtues of a

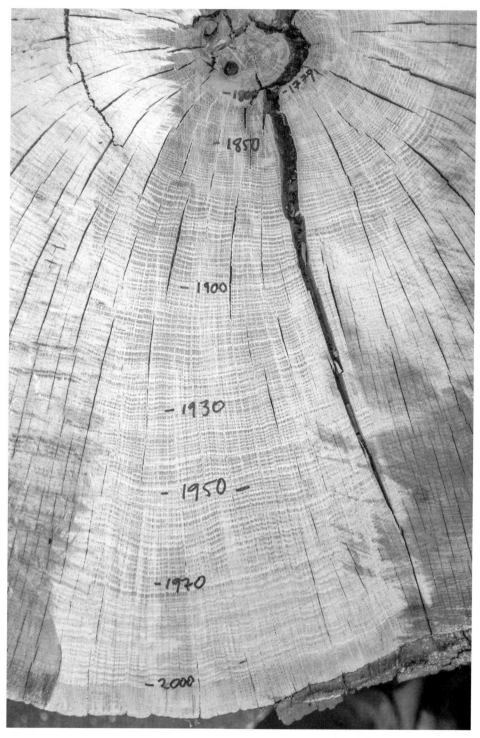

Figure 12.2. A bur oak "cookie" taken from a dead tree. This sample was from fourteen feet above ground and dates to at least 1620.

Figure 12.3. A couple of young biologists counting tree rings, an activity that kids find quite engaging.

five-hundred-year-old tree. The only proper answer to the question "How old is that tree?" is to say, "It looks pretty old."

Occasionally, we have an opportunity to get a sample from a very old tree and determine its age. We call a cross-section of such a tree a cookie. Cookies taken from the base of an old tree are usually hollow at the core, making it impossible to determine the exact age. As an alternative, we can cut up a dead tree and look for the lowest section of the tree that is solid to the core. We recently obtained such a cookie from our friend Ian Hoffman, owner of Big Beaver Tree Service. The cookie is from fourteen feet above the ground, meaning that counting the tree rings would only tell us the age of the tree when it was at least fourteen feet tall. After laborious sanding to make the rings visible, we counted the rings with the help of a couple of our young friends. The tree dates to at least 1620 at fourteen feet aboveground, but it may be much older—the inner rings show that this tree grew in shade, possibly the shade of giant cane, and the rings are so narrow that we had to count them with a microscope. We continue to collect cookies and hope eventually to develop a clearer picture of how old these trees are.

The other question about tree age that we commonly hear is "How much longer will that tree live?" To this there is again no answer. Trees are indeterminate organisms. That means that they are constantly building new structures out of stem cells. In a tree the stem cells are the apical meristem cells at the tip of each shoot and the cambium, a thin sheath of stem cells between the inner bark and the wood. Every year, a tree makes new leaves, shoots, flowers, and fruits from the apical stem cells, and new inner bark (phloem) and wood from the cambium. The same thing is taking place in the roots: apical stem cells constantly make new roots or elongate existing roots, while the cambial stem cells increase the diameter of the roots.

As long as this process of continued renewal is allowed to take place, the tree will not die. This way of growing differs from that of an animal, in which nearly all the stem cells differentiate into mature tissues. You can't grow a new limb if you lose one, but a tree can create new branches, new roots, even a new main stem if it is damaged.

Trees are very conservative in how they manage their population of stem cells. The vast majority of buds never grow out into stems. As the stem thickens, these buds get swallowed under the bark, but they continue to grow for the life of a tree. The stem of a tree contains thousands of these trace buds, little bundles of stem cells that sit under the bark and appear to do nothing. But when a tree loses its crown to lightning, wind, insects, or chain saws, those buds are released and grow to form a new crown. The Coldstream Tree is a good example of a tree that has lost its crown and released large numbers of stem cells to produce new branches. These vertical branches have built a new crown. Although the tree may look damaged, it is fully recovered from its injury and can live a very long life.

A tree that loses a limb, many limbs, or even its entire crown can continue to grow and thrive. This is very important to our ability to manage venerable trees. We must not assume that a tree with dead limbs or whose top is destroyed in a storm is a lost cause that needs to be cut down. There is no reason that such a tree cannot live out its normal life cycle as if it had never been injured.

The stem cells that build the tree are surrounded by the living tissues of the tree, including the xylem and phloem that conduct water and nutrients and provide support, the parenchyma cells that store nutrients, and other specialized cells. The living cells eventually die. Some dead cells are cast off in leaves, branches, and bark, but the wood cells remain a part of the tree for its whole life. A large tree is a scaffold of living cells built on top of its dead cells, and most of the mass of the tree is dead.

Many venerable trees show signs of decay. Some are quite hollow,

Figure 12.4. A blue ash with decay in an urban park. The bark has been killed, probably by lightning. The wood is decaying, but as long as the growth of the tree outpaces the growth of decay fungi, the tree will survive. Because it is in an urban park with high traffic, the tree should be carefully monitored.

Figure 12.5. A blue ash that died from unknown causes. The tree was allowed to stand in the Griffin Gate area because it would do little harm if it fell.

Figure 12.6. The blue ash pictured in figure 12.5. The tree fell after a few years, showing that it was extensively decayed. It was removed shortly after it fell.

others have the fruiting bodies of fungi popping out of their stems, and still others have decaying branches. Sometimes this can mean that the tree is doomed and that it needs to be removed or it will soon fall. But this is often not the case. If we took down every hollow or decaying tree, we would soon have nothing but barren grasslands.

I like to think of the relationship between a tree and decay as a competition. Decay is caused by a complex array of fungi, bacteria, insects, and other invertebrates, but it is the fungi that play the starring role. Decay fungi eat wood for a living. They might prefer to consume only cellulose or only lignin or both, but they convert wood to food. The process is slow. Cellulose and lignin, the main components of wood, are very hard to digest. And the tree is not defenseless. When a tree is wounded and invaded by the spores of fungi, living cells in the tree immediately go to work to create a chemical wall that makes it even harder for the fungi to digest the wood. And the heartwood, the dead parts of the stem, is filled with nasty chemicals that the tree makes to slow down the digestion of wood by fungi.

Once a fungus invades, the competition begins. The fungi consume wood to make fungal tissue as fast as possible, while the tree makes cells full of nasty stuff to slow down the fungus. As long as the tree's growth is equal to, or faster than, the growth of the fungus, the tree will survive. The tree

Figures 12.7, 12.8. A bur oak behind a house. This tree was alive, but after it was carefully assessed by Dave Leonard, an experienced arborist, the home owner chose to remove the tree. The consequences of a tree like this falling are obviously far greater than they are for the tree in Figures 12.4 and 12.5. A large tree's removal can represent a substantial cost: a large crane and a highly experienced crew of arborists were required to remove the tree without damage to the house.

may become quite hollow without losing a substantial amount of strength and without losing any growing tissues. If the growth of the tree slows down because of stress such as soil compaction, then the fungus can overwhelm the tree's defenses and the tree will die.

A general recommendation for the management of old trees is that they be allowed to stand without being removed, even if they are in advanced decay or are dead. The exception to this is when a tree would hit a target upon falling. A target might include a building, street, sidewalk, or heavily trafficked area in a park. Whether to leave a declining venerable tree in place depends on the consequences of the tree's falling down. A park with a playground presents a greater opportunity for damage, whereas a grazed woodland pasture presents a low risk.

There are exceptions to this conservative notion of management. Some fungi are not mild-mannered decay organisms but aggressive agents of disease. A well-qualified arborist should be consulted to advise a plan of management for an ancient tree with evidence of disease or decay. At a minimum, an arborist should be certified by the International Society of Arboriculture and hold advanced certification in tree risk assessment. Unfortunately, there are only a handful of arborists who have the advanced qualifications and experience to diagnose and care for venerable trees properly.

Once a tree is dead, its usefulness is not at an end. A dead tree, either standing or lying on the ground, is a valuable resource for wildlife, including birds and mammals, as well as beetles and other invertebrates and fungi. The dead tree will slowly decompose and enrich the soil.

Unfortunately, we often don't have the luxury of making a sound and rational decision about the fate of a tree. All too often, people with little understanding of the care of ancient trees remove them because of a perceived threat or because it is inconvenient to leave them. We lose all too many trees because of a lack of any professional evaluation of a particular tree or any serious discussion about its future. As we will see in the next chapter, public discussion of the fate of a tree and citizen involvement are critical to our ability to preserve ancient trees.

13

The Old Schoolhouse Oak

Extending the Lives of Old Trees

Not many trees make it onto the front page of their local paper, but the Old Schoolhouse Oak has done it many times. One of the largest bur oaks in the Bluegrass, and probably one of the oldest, is along the same road as the Ingleside Oak, the old buffalo trace. It is a more discreet tree, less apparent to passersby as it sits on a hill above the road. Until recently, few people were aware of the tree, but today the whole city knows it well.[1]

The Old Schoolhouse Oak is important not just because of its long life, but because of its repeated brushes with death. The tree was on the edge of a farm property in the middle of fairly new housing developments. In 2008 the property was slated for development, and the landowner intended to take the tree down. The tree appeared on the front page of the *Lexington Herald-Leader* when residents of the area objected loudly to the development plan. Their voices were heard: the planning commission objected to removal of the tree, and the developer agreed to reposition a road to save the tree. Shortly thereafter the housing market collapsed, and the development project was abandoned. In 2013 another developer began planning to turn the farm into homes and apartments in a manner that is consistent with the city's master development plan.

Local residents again raised vocal objections. The developer, Ball Homes, consulted with several experts, including me, to develop a preservation plan for the Old Schoolhouse Oak. During a careful examination of the

Figure 13.1. The Old Schoolhouse Oak at first inspection.

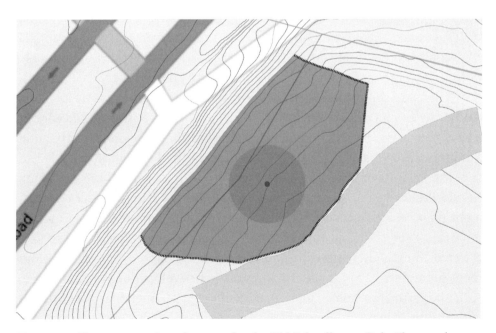

Figure 13.2. Tree preservation plan map for the Old Schoolhouse Oak. The tree (center dot) and its crown are surrounded by a three-quarter-acre mulched area (dark gray) and a construction fence (dashed line) to prevent encroachment.

Figure 13.3. The Old Schoolhouse Oak was carefully inspected and weak branches were removed.

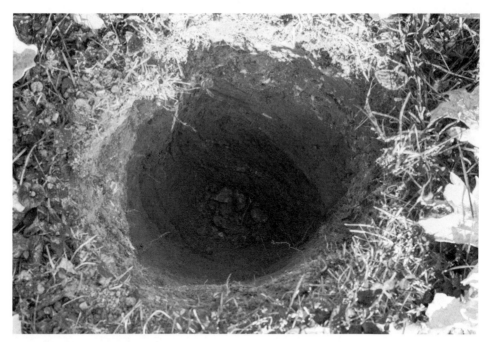

Figure 13.4. A soil pit near the Old Schoolhouse Oak, one of thirty that were used to assess soil quality and root distribution. In spite of the deep, fertile soil, there were no bur oak roots in any of the pits, which indicated that the tree was deeply rooted.

site and the tree, we realized that the tree had an extremely deep root system in a very deep and rich soil. It seemed likely that it could be preserved if given adequate space. The plan was to provide a large area, three-quarters of an acre, around the tree, along with mulch, a fence to prevent construction damage, and a long-term preservation document to plan for the future of the tree.

The preservation plan was carried out by Ian Hoffman of Big Beaver Tree Service over the next several months. The crew from Big Beaver pruned the tree to remove a few weak and damaged branches, cleared the dense tangle of shrubs and trees from around the Old Schoolhouse Oak, and laid down layers of kraft paper and mulch to prevent weeds from growing and avoid the use of herbicide.

All this time, the tree was surrounded by a chain-link fence on which were posted signs to warn construction contractors to stay away. Heavy equipment was forbidden to enter the tree-protection zone at any time. A common problem when construction takes place around trees is a failure to adequately protect the soil from compaction and the root system from injury. Even when protective fences are placed around a tree, contractors

Figure 13.5. The Old Schoolhouse Oak surrounded by its tree-preservation fence and signs to warn construction workers to stay away.

Figure 13.6. A different construction area in a public park where no tree protection zone was established and there was no contractual language to prevent contractors from compacting soil and damaging roots and stems. This probably will cause dieback and decline in this old Shumard oak.

Figure 13.7. The Old Schoolhouse Oak seen from the Military Pike. Freed of surrounding brush, the tree stands tall on top of its hill for all to see; it has become the icon of the new housing development.

often violate the protection zone. The only way to prevent this from happening is good fencing and enforceable provisions in contracts, followed by supervision and enforcement. In many cases there is no tree-protection language in construction contracts, or there is no on-site inspection and enforcement of contract provisions.

After the clearance of the brush and trees surrounding it, the huge tree on top of a hill suddenly became the most prominent feature of the landscape. The *Herald-Leader* received calls from people wondering where such a large tree had come from, and the tree was once again featured in the newspaper.

Can we guarantee that the tree will live for a long time? No, for the same reason that a physician cannot guarantee that a healthy but very elderly person will have a much longer life. We know very little about the management of elderly trees, and we don't know how much longer the Old Schoolhouse Oak will live. The best we can do is take the greatest care possible to minimize disturbance and stress and allow the tree to live out its natural live. We cannot stop such forces of nature as storms, and this is a very exposed tree. But we can prune properly to reduce wind loading, and we have already installed a lightning protection system.

The most important lesson from our experience with the Old Schoolhouse Oak and the St. Joe Oak is that citizen involvement is critical to the preservation of venerable trees. There are other trees that have received a last-minute reprieve from the bulldozer thanks to citizen awareness. One old bur oak was threatened by the construction of a strip mall and parking lot, but the developer agreed to preserve the tree and leave space for it. That tree sits in a tight little lawn next to the strip mall. It has been twenty years since the mall was built, but that does not mean the tree has survived the stress of confinement to a small space. The tree is in slow decline, and every year the signs of stress are more evident. The space is simply inadequate. But at least it is still alive.

The same cannot be said for one of the most magnificent ash trees I have ever seen. I came upon the tree while scouting around a new housing development for some left-behind trees. The tree was in the middle of an active land-clearance operation for a housing development. It appeared that the tree was to be left, like the Old Schoolhouse Oak, to become an icon of the new development. The contractor had cleared all around the tree. I was struck by how huge the tree was and took many photographs. A few days later, I came back to measure the tree, convinced that it might be a record tree, in diameter if not in height. But it was not to be: the tree was a pile of mulch.

I don't know whether this magnificent tree should have been preserved. It had a considerable amount of decay, but as we have seen, this is often not a serious issue for our long-lived trees. The problem is that there was, to my knowledge, no public discussion of the fate of the tree or any professional evaluation of the tree's condition. We need to make a commitment to public discussion of ancient trees in future developments.

We cannot preserve every tree, nor should we. By focusing development within current urban areas, including infill of small farms engulfed by urban development, we can preserve more of the rural environment that makes the Bluegrass and the Nashville Basin so attractive as a place to live and visit. Smart development will preserve as many of these old trees as possible.

We need to make conscious decisions about taking down venerable trees that are hundreds of years old. Too often, as happened to the huge ash that I found, the decision to remove a tree is made without any consideration or public discussion. A few years ago in Danville, Kentucky, maintenance workers felled the largest known blue ash tree. Their excuse was that they did not know it was important.

Figure 13.8. A magnificent blue ash tree on the edge of a development.

Figure 13.9. The same site a few days later. The tree was felled and ground into mulch without any professional evaluation of the tree or any rational decision about whether it would do well in the new development.

A venerable tree in a public place, such as the Old Schoolhouse Oak, will make it into the newspaper, and we can then have a public discussion about whether to preserve the tree or remove it. There are plenty of vigilant citizens who care about old trees and civic-minded developers and landowners who will try to do the right thing.

Private property rights are central to our society, and they should be infringed on as little as possible. Venerable trees, though, are a community resource with benefits that go far beyond property lines. They are a part of the history of our communities, and they provide measurable environmental benefits. They also improve property values: a neighborhood with beautiful, old trees is worth more than one without them.

Most of the trees that we lose never get the kind of attention given to the Old Schoolhouse Oak. There is no public discussion and no rational decision making if we don't know that a tree exists or is threatened with removal.

Conservation of venerable trees requires us to take a more active approach to their protection. An important rule of natural resource management is that we can't manage a resource until we know what it is. As a first step in managing our venerable trees, we should find out how many are left and exactly where they are.

Once we know where the trees are, and something about their condition, we can create better rules for making decisions about whether to keep them or let them be removed. In the case of the Old Schoolhouse Oak, it was possible to save one old bur oak but at the expense of other trees on the site, none of which was over one hundred years old, for there was not enough space to preserve all of them. The decision to preserve one tree at the expense of other, smaller trees was made by a consultative process among the developers, experts on old tree care, and local citizens. We cannot expect such a process to preserve every tree, but it will lead to better decision making. It is certainly true that, without strong community involvement, the Old Schoolhouse Oak would be gone.

Once the decision is made to keep a tree, or a grove of trees, a management plan is essential. We know very little about how best to manage venerable trees to allow them to live their full life span. The task is made more difficult because we don't know much about the biology of old trees. How large a root system do they need? Is pruning a good idea, or a bad one? If a tree has wounds or defects, are they too severe to allow the tree to be kept?

Though we don't know all the answers, we can develop some general rules of management.

Understand the tree. A careful evaluation of the tree by an expert is required. Examination of root distribution, growth rate, stem defects, and hazard potential will provide guidance for a preservation plan.

Create a permanent tree-protection zone. A tree-protection zone is usually designed to protect a tree during construction, but for venerable trees, that protection has to be permanent. That does not mean a permanent fence, but a permanent low-impact use area.

Give the tree the maximum possible space. There are agreed-on standards for how much room a tree needs, but those are minimums. It is reasonable to suggest that a venerable tree deserves a half acre (two thousand square meters) or more, and a grove of trees needs proportionally more. That does not mean that the area needs to be symmetrical around the stem.

Give the tree maximum protection from construction. Construction fences designed to protect a tree are often violated by careless contractors. Contractors should be given absolute instructions not to violate the sanctity of a protective fence.

Minimize competition from other plants. Especially during construction, soil surrounding a tree should be covered in mulch to eliminate grass and other plants.

Avoid the use of herbicides or fertilizer. It is a common error to think that a stressed tree needs to be fertilized. The opposite is probably true—fertilizer is itself a stress factor for old trees.

Minimize lawn care. After construction, a lawn can be established around the tree if necessary, but it should not be managed as a pristine, high-input carpet of pure grass. Such a lawn requires herbicides and fertilizer that are not good for the tree.

Avoid high-impact recreation or other uses that cause soil compaction. A playground or ball field is not an appropriate use of the tree-protection zone. Low-impact use, such as a picnic area, is less likely to harm the tree.

Allow natural nutrient cycling. Trees are experts at recycling. When leaves, fruit, and twigs fall and decay, their nutrients are recycled into the soil by complex interactions among the tree, microbes, and invertebrates. Leaving a large mulched zone around the tree and allowing the leaves to accumulate in the fall and decay naturally will reduce the need for fertilization.

Monitor the tree. The work is not done when construction is complete and the property is sold. The new property owners and managers, including home owners' associations, should be provided with detailed documentation about the tree, its history, and its need for care. The tree should be regularly inspected by a qualified professional, pruned as needed to avoid the fall of large dead branches, and assessed periodically for potential risks.

Eventually, all these trees will die, for as incredibly long-lived as they may be, they are mortal. We need to plan for their replacement and management, and on this we are doing a terrible job in all our cities. For the most part, we are planting the wrong trees in the wrong places. Our venerable tree species need space—as we have seen, they are not suitable as street trees or in small yards.

We plant the wrong trees because the right trees are not generally available. The modern horticulture industry provides us with huge numbers of cheap trees made from cuttings rather than seeds. All the trees from cuttings of a single variety of tree form a clone—a group of genetically identical individuals. Each of the trees in the clone is called a ramet. Planting ramets reduces urban tree diversity—sometimes an entire neighborhood is planted with a single clone of red maple, so there is no genetic variation. The loss of American elm was a devastating blow to many cities, and the lesson we should have learned was that we needed to promote a high diversity of tree species in urban areas—and genetic diversity within species. Instead, we have done the opposite by planting huge numbers of short-lived, cloned trees with limited genetic variability. (See chapter 14 for additional discussion of reproduction through cloning.)

Even if we have a large enough planting space for our venerable tree species, finding seedlings to plant is difficult. Bur oak is readily available, and many, perhaps most, bur oak seedlings in the nursery trade come from seeds collected in Lexington. Shumard oak seedlings and cultivars are readily available and commonly planted, but the seed source for most of these is from the Deep South (they are commonly called Texas Shumard oak), and they may not be suited to a long life in our soils. Chinkapin oak is available from a few nurseries, whereas kingnut and blue ash are rarely available. Many nurseries are afraid to plant blue ash because of the potential for damage from emerald ash borer. Kingnut, like many hickories, is difficult to transplant. Careful site selection is critical, and we earlier discussed appropriate landscapes, such as church and corporate campuses.

The best solution to the lack of availability of seedlings of our native tree species is to develop local nurseries that plant locally collected seeds and avoid the use of clones. This is unlikely to be a highly profitable business, however, and probably needs to be a part-time occupation for a farm.

We have been fortunate enough to have these venerable trees in our midst for as long as our towns and cities have existed. It is up to us to determine whether they are here hundreds of years from now. We can start by being mindful of their presence, by appreciating their age and beauty, and by valuing their contributions to our lives. But we need to take action as well, not only to preserve existing trees in our urban and suburban landscapes, but to plant long-lived trees wherever it is appropriate.

14

The Future of Venerable Trees and Woodland Pastures

The woodland pastures of the Bluegrass and Nashville Basin are a legacy created by nature and left to us by our ancestors. The woodland pastures and their venerable trees have been with us for hundreds of years. They will not be with us much longer if we do not take action. We have to decide whether the trees that make the Bluegrass and Nashville Basin such special places in which to live and farm will be here for future generations. A drive through Bourbon County, Kentucky, shows our two alternative futures—one that is still graced by ancient trees in woodland pastures, and one that is devoid of trees.

As we have seen, the trees are getting old, and both poor management and development are hastening their demise. This would not be a problem if they were being replaced. Like any organism, trees have a finite lifespan, even if it is a very long one. Sooner or later, every tree dies. Any tree that fails to reproduce will eventually disappear. As our ancient trees die, our landscape will become increasingly barren.

Some landowners may think they are doing the right thing by planting large numbers of trees. It is true that many farms, especially our famous horse farms, have a lot of young trees. The problem is that they are often the wrong trees—they will not ensure a future of woodland pastures. The trees being planted today are fast-growing, short-lived species such as red maple and Norway spruce. These trees may fill a vacancy in the landscape for a short while, but they will be gone in a few years and will never reach the size and stature of our ancient woodland pasture trees. In a few cases,

Figures 14.1, 14.2. Which future do we want for our farmland? Bourbon County, Kentucky.

The Future of Venerable Trees and Woodland Pastures   191

Figure 14.3. A former woodland pasture, part of the Coldstream complex, was replaced by modern, short-lived ornamental trees.

trees such as northern red oak or baldcypress may reach sufficient size and age that they will be with us for a long time, but for the most part we are replacing very long-lived trees with short-lived ones.

To understand why we are making such poor planting choices, it is helpful to know a little bit about the nursery industry that provides our planting stock. The nursery trade in the United States began with local nurseries producing tree seedlings, often from locally obtained seeds. The first nurseries, dating back at least to 1640, were devoted to production of fruit trees, especially apples. As more of the New World was explored, and our astonishing diversity of plants was discovered by explorers and botanists, a market grew for the export of trees and shrubs to Europe. Nurseries along the Atlantic coast, especially on Long Island, were producing seedlings of almost every American tree. As the nursery trade grew, it increasingly became concentrated in upstate New York. By 1850 the largest tree nursery in the world, Ellwanger and Barry Nursery of Rochester, was shipping tree seedlings all over the world and sending expeditions out to find new plants. A thriving world trade in trees developed: American trees showed up in European and Japanese gardens, and Eurasian trees were planted in the United States and Canada.

Luther Burbank, the great self-educated pioneer of modern horticulture, further transformed the industry when he began selecting and breeding new ornamental and food plants, introducing more than eight hundred new varieties of plants. Burbank was an inventor of new plants in the same way that Thomas Edison was an inventor of machines. But Edison had an advantage that Burbank could not enjoy, the power of the patent. Edison, like other inventors, could register a patent for his inventions that would prevent, or at least discourage, others from copying his work. Plants, though, could not be patented, so Burbank had no way of preventing other nurserymen from taking cuttings of his plants and selling them as their own.

It was not until 1930, four years after Burbank's death in 1926, that the horticulture industry began enjoying the same protection that Edison had. Burbank bequeathed most of his varieties to Paul Stark, his friend and fellow plantsman. Stark, who operated a family-owned nursery in Missouri, realized that he had a treasure trove in the seeds and plants that Burbank had bequeathed him, but that without the ability to patent the plants, his competitors would soon be making copies of his plants. It was the leadership of Stark, the legacy of Burbank, and the championing of Stark's efforts by Thomas Edison that led Congress to pass the Plant Patent Act of 1930. This gave plant breeders the same protection of their intellectual property that manufacturers had long enjoyed.

Not all plants could be patented. Congress decided that only plants that could be reproduced from cuttings could be patented, not those produced from seed. This made sense at the time—seeds are produced by genetic recombination, and seedlings are not genetic copies of the parent plant. A cutting of a plant can be used to make endless, genetically identical copies.[1]

Patent protection had the desired effect. By reducing competition, it allowed the horticulture industry to grow and thrive. Stark Bros. became one of the most successful tree nurseries, and it remains so to this day. We have all benefited from Paul Stark's work to protect Luther Burbank's legacy. The astonishing array of plants in our gardens and landscapes would not have been possible if breeders had not been able to protect their work.

What was good for the nursery business may not have been so good for the trees in our landscape. The plant patent system led the horticulture trade to largely abandon the production of tree seedlings in favor of endless copies of the same plant, creating a clone of genetically identical individuals. Because of the plant patent law, clones make money and seedlings don't. This is narrowing the genetic basis of our landscape. A horse farm

The Future of Venerable Trees and Woodland Pastures   193

Figure 14.4. A uniform row of genetically identical "Autumn Purple" ash trees.

that plants fifty red maples of a single variety is not planting fifty different trees, but fifty copies of the same tree.

There may be other disadvantages to producing plants from cuttings, especially when they are grafted onto roots of a different tree. Most trees that are sold in the ornamental trade have not been tested for long periods, and problems that do not show up in the nursery may limit the life of the tree in the landscape. The "Autumn Purple" white ash that was so popular for a few years is grafted to green ash rootstock. Over the long run, the graft proved to be incompatible, and the trees often died after only a few years in the landscape.[2]

Modern horticulture is designed to put beautiful plants in the landscape. It is not designed to keep them there. The combination of weak, genetically uniform plants with high inputs of fertilizer, pesticide, and herbicide and stress from mowers, string trimmers, and heavy equipment is the antithesis of the kind of low-input sustainable landscape management that leads to long lives for trees. Many experienced arborists also believe that the long-term survival of containerized plants, which is what you will find in most nurseries, is less than the survival of bare-root or balled-and-burlapped plants. Direct seeding, whereby the roots are never disturbed, may in the long run prove to be the best way to start trees off to a long life.

Forest tree nurseries have taken the opposite approach from that of ornamental nurseries. A forest tree nursery tries to get the most genetic diversity possible by collecting seeds from multiple sources within a defined distance of the nursery. As a result, seedlings from a forest tree nursery will always have greater genetic diversity than clones from an ornamental nursery.[3]

The long-term sustainability of our woodland pastures will depend on our ability to grow trees with high genetic diversity, using the methods of forestry rather than horticulture to maximize longevity and genetic diversity. We need to plant seedlings, not clones.

Creating Sustainable Woodland Pasture Landscapes

The woodland pastures and our venerable trees are the products of a naturally occurring system of drought, karst, and heavy, intermittent grazing. Today we still have intermittent drought and karst, but we have shifted from heavy intermittent grazing to continuous low- to moderate-intensity grazing (horses at the low end, cattle at the moderate). The closest we probably come to the original system is rotational intensive grazing such as what occurs at Elmwood Stock Farm. This provides the complete grazing of all the forage plants, some churning of the soil by hooves, and a recovery period while the cattle are moved off. It does not provide long enough periods between grazing to allow trees to regenerate. We lack the one key ingredient to make our landscape self-sustaining—a grazing animal that leaves for long periods.

There are two ways we can create a sustainable woodland pasture landscape. Either we can introduce intermittent grazing, or we can replace the natural regeneration process with an artificial one by planting seedlings of venerable tree species.

Allowing livestock to graze only every five to ten years does work. Jane Julian's fine stand of old woodland pasture trees shades a thriving population of young trees because she allows the cattle into her woodland pasture only at long intervals. For a nature preserve or public land, this is a viable solution. Griffith Woods Wildlife Management Area uses herbicides, mowing, and fire in a bootless attempt to foster the growth of desired trees, when simply allowing cattle to graze the woodland pasture at long intervals might foster the growth of the replacement trees. In fact, the abandonment of grazing at Griffith Woods is probably what allows the mother tree to reproduce, as we saw in chapter 7. It is probably not realistic to reintroduce bison to this area because the amount of land available for them to wander is not sufficient.

Most of our woodland pastures are found on farms that are too productive of horses and cattle to allow long grazing intervals. This is, after all, some of the best pastureland in the world.

We can grow and transplant seedlings of venerable tree species. Bur oak is readily available from many nurseries, and these are usually seedlings, not cloned from cuttings. Chinkapin oak is less readily available, but a few nurseries carry it. Kingnut and blue ash are largely absent from either horticultural or forest nurseries—they are difficult to establish and slow-growing. Kingnut is very hard to transplant, and its failure rate is high. Shumard oak can be obtained either as cultivars of rooted or grafted cuttings or as seedlings. As we discussed earlier, however, there is reason to think that the Shumard oak of the horticulture trade and the local native trees are different.

Forest tree nurseries, such as the two nurseries operated by the Kentucky Division of Forestry, produce seedlings from locally collected seeds. Forest nurseries produce small seedlings for reforestation, typically selling one-year-old trees in bulk. Larger trees are more desirable for landscape purposes, but few nurseries are currently producing large tree seedlings.

THE TREE PEN

Tree pens are common on farms in the Bluegrass. They are areas of a pasture or paddock encircled by a fence, usually the black plank fences common in this area. Sometimes a tree pen encloses one or more huge old trees, but often is used to plant groups of smaller trees. The tree pen might be an ideal context in which to regenerate woodland pastures either by planting or by allowing natural regeneration.

Toss Chandler is an artist and lover of trees who has carefully tended the magnificent trees on the farm that has been in her family since Kentucky was the wilderness. A few years ago, she built a pen around an ancient, doddering bur oak to protect it, as she put it, in its last years. The tree has since died, but not before dropping a lot of acorns that came up in the shade of the mother tree. Today in this pen is a little stand of bur oaks, about fifteen trees.

Natural regeneration is probably not reliable enough. Toss was fortunate that she had just the right combination of shade to keep down competing trees, and then the death of the mother tree to release the seedlings.

Direct seeding shows promise for tree pens. Kingnut is very difficult to transplant, but as we have seen, the mother tree at Griffith Woods has produced abundant progeny. Direct seeding of a mixture of woodland pasture trees is worth trying, although weed control and predation by squirrels may limit success. One promising method is direct seeding into tree shelters, which protect the young seedlings from stress and predators while allowing them to develop robust root systems.

Some landowners will not want to wait for seeds to grow into big trees, but will prefer to purchase larger material. We need more local nurseries that can grow trees from locally collected sources. Rather than establishing a large and expensive commercial nursery, it makes more sense to start a network of small cooperative nurseries on existing farms from locally collected seeds. Low-maintenance, organic nurseries that avoid the use of herbicides and excessive fertilizer will produce hardy planting stock that can supplement natural regeneration.

Once trees are established, they need to be managed properly. The space inside a tree pen can be a management problem. In the absence of grazing, the pen is likely to fill with honeysuckle or other undesirable plants. Periodic grazing inside the tree pen is ideal, but mowing may be required if grazing is not practical. Pastures managed for cattle using intensive rotational grazing do not require fencing to protect mature trees, but tree pens are useful for tree reproduction. One solution to the problem of weeds in the tree pen may be to plant cane, the natural companion of the venerable trees. Cane is difficult to transplant, but once it is established it is self-perpetuating.

Livestock will occasionally chew the bark of older trees. Although they rub on trees and do a bit of chewing, horses usually will not break through the protective barrier of the outer bark. But occasionally horses will thoroughly chew the bark and kill a branch or tree. This can be prevented only by separating the horses from the trees with a fence. Cattle are more likely to chew on bark, and there generally should be a fence between livestock and trees.

The long-term care of a tree should be minimal. In a tree pen, leaves that drop in the autumn can be allowed to decay, which minimizes the need for fertilizer. Watering may be required in the first couple of years of establishment of new trees in the event of a drought, but regular watering after that should not be necessary. Pruning of dead branches should be avoided as long as the branches will not do serious damage should they fall. A dead branch is an important wildlife habitat.

THE PRESERVATION CHALLENGE

There are some who argue that we need to "restore" the original Bluegrass woodland pasture habitat, including not only the trees but the native grasses, herbs, and shrubs. This has been attempted at Griffith Woods and a few other places, with a notable lack of success. The original herbaceous plants of the Bluegrass have been gone for centuries, and they are largely unknown. As we have seen, the first settlers quickly converted the woodland pastures created by bison to woodland pastures grazed by cattle and horses. As early as 1800, farmers planted nonnative bluegrass, fescue, and clover, replacing the native grasses. The cane, which once covered enough land to allow Josiah Collins to become lost for several days, is now nearly gone, existing only in a few patches here and there. Griffith Woods and some other preserves where grazing has been excluded are rapidly being converted to forests. The presence of aggressive nonnative weeds like bush honeysuckle restricts the reproduction of desirable tree species, and the forests that replace the woodland pastures are of poor quality.

There is certainly a need to provide habitat for the herbaceous species that we know were in the Bluegrass. A prime example is running buffalo clover (*Trifolium stoloniferum*), on the federal endangered species list. As the name implies, this is a species that was as dependent on disturbance by bison as our woodland pastures. In the absence of intermittent grazing, and with the increasing presence of nonnative plants such as bush honeysuckle, it is difficult to see how this species can recover.

As we have seen, the woodland pasture habitat was created by grazing animals. If not for drought and bison, the woodland pasture habitat would never have existed. Without a large grazing mammal—whether bison, cattle, or horses—any woodland pasture will quickly become a forest. In the thirty years that I have been visiting Griffith Woods, the entire appearance of the woodland pastures has changed as native and nonnative trees and shrubs move in; it is rapidly becoming a forest. In contrast, Julian Farm, where cattle are allowed in to graze at intervals, is maintaining a healthy population of woodland pasture species young and old.

It is futile to set aside land to try to restore the Bluegrass to some imagined past without large mammals, and we should be very reluctant to take more woodland pastures out of production. Our woodland pastures are very valuable for livestock production, and our venerable trees can peacefully coexist with livestock as long as we take some simple measures, such as creating tree pens and planting appropriate replacement trees.

VENERABLE TREES AND CLIMATE CHANGE

Venerable trees face many challenges and deserve our best efforts to prolong their lives and ensure their futures. The single biggest threat to the continued existence of these trees is global warming or climate change.

Although there is a lot of natural variation in climate and weather, we have known for more than one hundred years that burning fossil fuels is increasing the concentration of carbon dioxide in the atmosphere, which is causing the world to become warmer. Carbon dioxide and other gases act like a blanket over the earth, trapping the sun's heat. The term *global warming* implies that the effects of adding greenhouse gases to the atmosphere are worldwide, and that is true. The consequences of climate change, however, are not uniform—some regions are warming faster than others, some are becoming drier and some wetter. In recent years, as the science of global warming has improved, our ability to predict regional effects has also improved.[4]

Climate change could make conditions in the Bluegrass and Nashville Basin dramatically different for our native trees. It may seem that we can't predict the effects of a warmer world. After all, it is difficult enough to predict next week's weather. In fact, though, scientists have been extraordinarily successful at predicting the effects of global warming. If anything, we have underestimated the speed of change.

At present, we can say the following about the consequences of climate change in the Bluegrass and Nashville Basin. Summers will be hotter, both day and night. The frequency of serious air pollution, particularly ozone, will increase. Winter is less predictable. Although winters in general will be warmer, recent research has shown that the rapid warming of the Arctic Ocean can drive cold air streams into our region in winter. Rainfall will increase, perhaps by a lot, and more of our rain will come in major rainstorms—this is already happening. In spite of greater average rainfall, we are likely to experience more frequent and more severe droughts.[5]

What does this mean for our woodland pasture trees? All these changes, especially higher nighttime temperatures, higher ozone levels, and more frequent drought, are potential stress factors. We have already seen, though, that our woodland pasture trees are deeply rooted and extraordinarily drought-tolerant.

Some analysis is possible about the future suitability of the Bluegrass and Nashville Basin as a habitat for our woodland pasture species. Every tree lives within what we can consider an "envelope" of suitable conditions—enough water but not too much, warm or cool summers, soil depth and fertility. Every tree species is thought to have a different envelope in which it can grow and reproduce—the limits of temperature, soil properties, and moisture that determine tree survival. That is why we don't have palm trees in Kentucky or bur oaks in Florida. The conditions for tree growth at a single location change over time. When I first came to Kentucky more than thirty years ago, loblolly pine could barely survive here. Now there are loblolly pine stands thriving and even reproducing in western Kentucky. In Lexington and other communities, planted loblolly pines are growing well. Some of this is breeding—geneticists have been selecting loblolly pine trees for cold tolerance for over fifty years—and some of it is climate change—winters' low temperatures are not as extreme as they used to be.

Louis Iverson, Anantha Prasad, and their colleagues at the USDA Forest Service have been looking at the envelope of tree growth and comparing it with models for future climate. Their work looks at the potential suitable habitat for a tree species now and in the year 2100. We show some of their work here, but with a strong warning—these are predictions not about where a tree species will be found in the year 2100, but about where suitable habitat for that species might exist. A tree might never be found in a future habitat in 2100 if it has no way of getting there, and as is true of any modeling project, there is a lot of uncertainty. We will use their work as a guide to what might happen in the near future, well within the lifetimes of most trees.[6]

We begin with loblolly pine because it illustrates the dramatic effect of climate change on a commercially important tree species. Loblolly pine will probably be able to expand its habitat all the way to Maine, even as it continues to grow well in the South. Of course, it will grow in those northern habitats only if we plant it there. This analysis shows how dramatically the suitable habitat for trees can change with even moderate changes in climate.

Sugar maple grows very well in the Bluegrass and, as we have seen, was part of the original woodland pasture vegetation. The maps show that the Bluegrass and Nashville Basin will soon become much less suitable habitat for sugar maple, and its population will decline. I believe that this may already be occurring, as a lot of maples have succumbed to various ailments, although some of that is due to poor management rather than climate change. Black maple is not shown in this analysis, but it is more

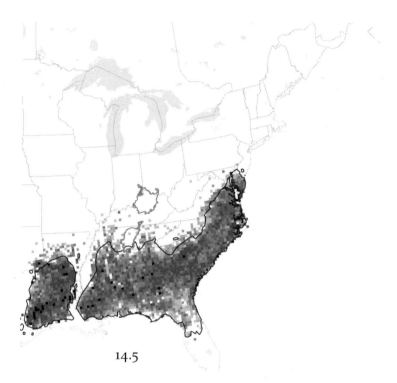

14.5

Figures 14.5, 14.6. Current (14.5) and future (14.6) habitat for loblolly pine. The black outline indicates the historical range. Figure 14.5 shows the current relative abundance (importance value), and figure 14.6 shows the potential future habitat.

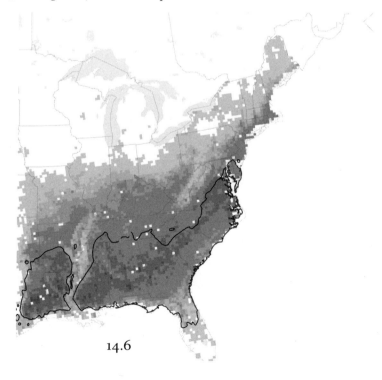

14.6

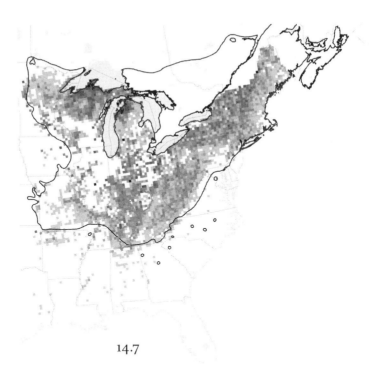

14.7

Figures 14.7, 14.8. Current (14.7) and future (14.8) habitat for sugar maple. The black outline indicates the historical range. Although sugar maple currently grows well in the Bluegrass, this analysis shows that the Bluegrass and Nashville Basin may not be suitable for sugar maple in the near future.

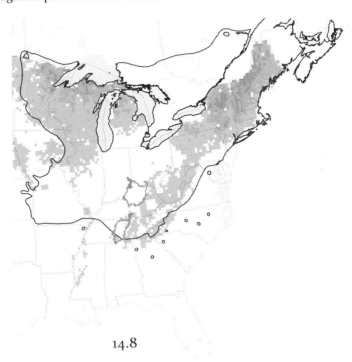

14.8

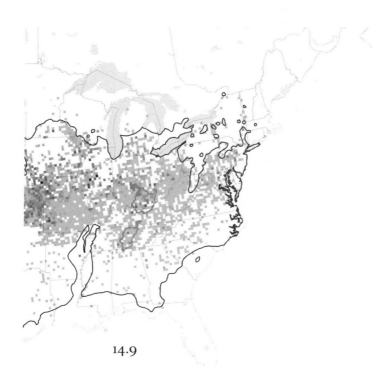

14.9

Figures 14.9, 14.10. Current (14.9) and future (14.10) habitat for black walnut. The black outline indicates the historical range. Black walnut is common in the Bluegrass and Nashville Basin, but likely to become less so in the Bluegrass and may not grow in the Nashville Basin.

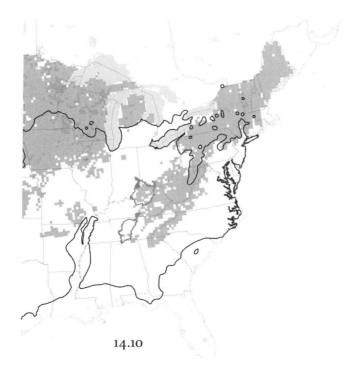

14.10

drought-tolerant than sugar maple. If we still want maples in our woodland pastures, we can either use black maple or begin selecting and breeding for more heat- and drought-tolerant sugar maples.

Black walnut is an abundant species in the Bluegrass and Nashville Basin. Suitable habitat will be farther north in the near future, and it may no longer reproduce or grow well in the Nashville Basin. The threat of thousand cankers disease may make the climate issues for black walnut more complicated.

The three predominant oaks of the Bluegrass and Nashville Basin, bur oak, chinkapin oak, and Shumard oak, are likely to do even better here in the future under the current trajectory of climate change. They will become more northerly species, but they will also expand in the Nashville Basin

We do not have enough data on kingnut or blue ash to be confident about future habitat suitability. As is the case with the other venerable tree species, their great drought tolerance is likely to allow them to tolerate a drier, warmer future.

The toughness of our venerable tree species, their ability to root deeply into rock, their extraordinary tolerance of drought all suggest that these trees will still be suitable for our climate into the future. We will lose some species completely, especially with the combination of climate change and introduced pests and pathogens. White ash is likely to disappear, and black walnut is at great risk.

Our trees could make a great contribution to assisted migration projects. Assisted migration is the process of moving species to places where future habitat will be more suitable. It is necessary because, unlike past changes in climate, the current global warming is happening too fast for trees to migrate on their own. We will need, very soon, to begin planting trees in suitable future habitats. Our very drought-tolerant trees, such as the bur oaks in the southern Nashville Basin, may become important seed stock for future tree-planting programs farther north.

Climate change is just one factor affecting the future of our woodland pasture habitat. We need to actively manage the trees of our woodland pastures to ensure their future. The choice of whether our ancient trees will be familiar to our descendants hundreds of years from now is entirely up to us. We can begin to act now, or we can sit back and watch our heritage disappear.

14.11

Figures 14.11, 14.12. Current (14.11) and future (14.12) habitat for bur oak. The black outline indicates the historical range. Bur oak may dramatically expand its range, owing largely to its drought tolerance.

14.12

14.13

Figures 14.13, 14.14. Current (14.13) and future (14.14) habitat for chinkapin oak. The black outline indicates the historical range. Chinkapin oak will continue to do well in the Bluegrass and Nashville Basin.

14.14

14.15

Figures 14.15, 14.16. Current (14.15) and future (14.16) habitat for Shumard oak. The black outline indicates the historical range. Shumard oak will continue to grow well in the Bluegrass and Nashville Basin.

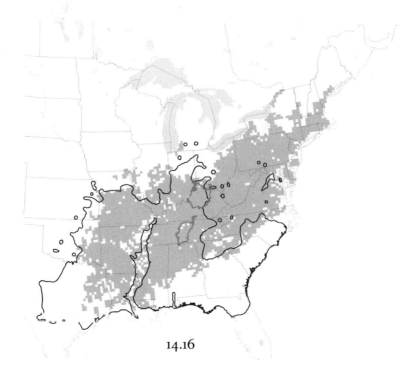

14.16

Acknowledgments

This book is the result of more than forty years of experience with trees, including thirty-two years observing the woodland pastures of the Bluegrass. Along the way, I have been advised, influenced, cajoled, and coaxed by many other scientists, teachers, colleagues, and students.

I owe a deep debt of gratitude to my alma mater, the State University of New York College of Environmental Science and Forestry, Syracuse. My teachers there, including Hugh Wilcox, Ed Ketchledge, Jim Geis, Milton Silverstein, Wilfred A. Côté Jr., Dudley Raynal, and others turned an eager young student into a scientist and naturalist. I am also grateful to the string of green buses that carried us from behind Moon Library into the magic world of field biology. It was that immersion in nature that made me a good enough observer to begin to understand venerable trees. At the University of Wisconsin–Madison I was taught by masters of forestry and botany, including Ted Kozlowski, Ray F. Evert, Ray Guries, Eldon Newcomb, and Folke Skoog. I am sure I have left out many other important teachers, for which I apologize. I also benefited greatly from the National Outdoor Leadership School's Outdoor Educators Course.

Over the years, I spent many happy hours in the field or talking about trees with friends, colleagues, and students, including Kathleen E. Moore, Judith Henningson, Bob MacDonald, Jim Dunn, Jeff Stringer, Hilary H. B. Smith, Catherine Monzingo, and many, many others. Robin W. Kimmerer and I learned field biology together—she looked down at the mosses while I looked up at the trees. I can only hope that some of her immense skill as a writer and observer of the natural world rubbed off on me.

Many people in Kentucky have helped me understand the venerable trees and the history of Kentucky. Larry Redmon is a font of knowledge about the early history of the region, as well as a lover of trees. B. J. Gooch, archivist at Transylvania University, and Ruth Bryan, archivist at the

University of Kentucky, were generous with their time in helping me search for historic documents. Many landowners provided access to their properties, and Jason DeBold has been my companion on many excursions around the Bluegrass in search of trees. Tim Diachun has helped me keep my feet on the ground, and been a good friend throughout. The many enthusiastic participants in the venerable trees workshops have been very helpful. Master Arborist Dave Leonard has been my friend and colleague for over thirty years and has contributed greatly to my understanding of the management of old trees. Ian Hoffman, of Big Beaver Tree Service, has worked closely with me on the project described in chapter 13, and I am grateful for his support. I thank Joyce Bender and the Kentucky State Nature Preserves Commission; the staff of the Kentucky Division of Forestry and their director, Leah Williamson MacSwords; the Kentucky Division of Wildlife; and the U.S. Forest Service for generous assistance.

The Carnegie Center for Literacy and Learning was my writing home for most of this book, providing me with writing space, the companionship and advice of other writers, and workshops to hone my skills. I am grateful to Neil Chethik, the director; to the fine writers in my nonfiction writing group; and to Nikky Finney for sage advice. I have also made use of space provided by Awesome Inc., and am grateful for the friendship of the awesome people there. A Cup of Commonwealth was my daily source of great coffee and good conversation while I worked on the book.

Many colleagues all over the world contributed to my understanding of venerable trees. I am particularly grateful to Tibor Hartel, Julian Hight, David Burg, Rob McBride, Frans Vera, Ted Green, Oliver Rackham, Ryan McEwan, Neil Pederson, Jill Butler, the members of the Urban Tree Growth and Longevity Working Group, and many others who generously provided their time and expertise.

Erin Barnhill edited the draft and turned it into something readable. Ashley Runyon, Ila McEntire, and Ann Twombly, my editors at the University Press of Kentucky, were wonderful to work with throughout the publication process.

Andy Mead deserves special thanks both for carefully reviewing the manuscript and for his many years of reporting on ancient trees while he was at the *Lexington Herald-Leader*. Tom Eblen of the *Herald-Leader* has done excellent reporting on our ancient trees, and he also has given me considerable guidance on the history of Lexington.

The photographs in this book are by the author, except as otherwise acknowledged. I thank Tibor Hartel, Julian Hight, and the State Tretyakov Museum of Moscow for the use of their images. I am grateful to Wilfred A. Côté Jr. for teaching me the skills of a scientific photographer, to Winston Hall for helping me improve my photography skills, and to Tony Kuyper and Sean Bagshaw for helping me learn digital post-processing.

In spite of all this help, I am sure there are errors of fact or interpretation in this book, for which I bear sole responsibility.

Appendix

Trees of the Bluegrass and Woodland Pastures

Taxonomy follows USDA with minor exceptions. Common names follow USDA Forest Service standard use. Presence in the Bluegrass is from the USDA plants database (http://plants.usda.gov); frequency information is from my experience. Presence in Kentucky woodland pastures is based on my experience and discussions with others, and "no" means that I have not seen it in woodland pastures, not that it is never found there. Shrubs are included if they are occasionally tree size. "Planted" means the species is planted but not known to be naturalized. "Naturalized" means that there are sexually reproducing populations of nonnative species.

Common name	Species	Family	Bluegrass	Woodland Pastures
ash, blue	Fraxinus quadrangulata	Oleaceae	abundant in dry woods	abundant
ash, green	Fraxinus pennsylvanica	Oleaceae	common along rivers, declining	no
ash, white	Fraxinus americana	Oleaceae	abundant, declining	abundant, declining
asper., bigtooth	Populus grandidentata	Salicaceae	uncommon	no
autumn-olive	Elaeagnus umbellata	Elaeagnaceae	occasional in woods near the Knobs, naturalized	no
baldcypress	Taxodium distichum	Cupressaceae	uncommon along Kentucky River, introduced elsewhere	no
basswood, American	Tilia americana	Tiliaceae	common	uncommon
beech, American	Fagus grandifolia	Fagaceae	common along rivers, lower slopes	no
birch, river	Betula nigra	Betulaceae	uncommon, along Kentucky River	no
blackgum	Nyssa sylvatica	Cornaceae	uncommon	no
blackhaw, rusty	Viburnum rufidulum	Caprifoliaceae	common in dry woods	occasional
blackhaw, Southern	Viburnum prunifolium	Caprifoliaceae	common in dry woods	occasional
bladdernut	Staphylea trifolia	Staphyleaceae	common near creeks	no
boxelder	Acer negundo	Aceraceae	common along rivers, creeks	occasional
buckeye, Ohio	Aesculus glabra	Hippocastanaceae	abundant	abundant
buckeye, red	Aesculus pavia	Hippocastanaceae	reported in Fayette County	no
buckeye, yellow	Aesculus flava	Hippocastanaceae	common along the Kentucky River	no
buck-horn, Carolina	Frangula caroliniana	Rhamnaceae	common in moist woods	no
burningbush	Euonymus atropurpureus	Celastraceae	common in woods	no
butternut	Juglans cinerea	Juglandaceae	rare, declining	rare
catalpa, northern	Catalpa speciosa	Bignoniaceae	common, naturalized	common, naturalized
cherry, black	Prunus serotina	Rosaceae	abundant	abundant
chokecherry	Prunus virginiana	Rosaceae	occasional	no
coffeetree, Kentucky	Gymnocladus dioicus	Fabaceae	common	common
cottonwood, eastern	Populus deltoides	Salicaceae	common along Kentucky, Salt rivers	no

crab, sweet	*Malus coronaria*	Rosaceae	uncommon	no
dogwood, alternate-leaf	*Cornus alternifolia*	Cornaceae	common	no
dogwood, flowering	*Cornus florida*	Cornaceae	common	no
dogwood, rough-leaf	*Cornus drummondii*	Cornaceae	common	no
elderberry, American black	*Sambucus nigra* ssp. *canadensis*	Caprifoliaceae	common along woods margins	no
elm, American	*Ulmus americana*	Ulmaceae	common	occasional
elm, rock	*Ulmus thomasii*	Ulmaceae	uncommon	occasional
elm, slippery	*Ulmus rubra*	Ulmaceae	common	occasional
elm, winged	*Ulmus alata*	Ulmaceae	occasional along Kentucky, Salt rivers	no
euonymus, winged	*Euonymus alata*	Celastraceae	introduced, common in woods	no
hackberry, common	*Celtis occidentalis*	Ulmaceae	abundant	abundant
hackberry, dwarf	*Celtis tenuifolia*	Ulmaceae	common	occasional
hawthorn, cockspur	*Crataegus crus-galli*	Rosaceae	common	no
hawthorn, downy	*Crataegus mollis*	Rosaceae	common	no
hawthorn, waxyfruit	*Crataegus pruinosa*	Rosaceae	uncommon	no
hazelnut, American	*Corylus americana*	Betulaceae	occasional along creeks	no
hemlock, eastern	*Tsuga canadensis*	Pinaceae	occasional in woods near the Knobs	no
hickory, bitternut	*Carya cordiformis*	Juglandaceae	abundant	common
hickory, mockernut	*Carya tomentosa*	Juglandaceae	common	occasional
hickory, pignut	*Carya glabra*	Juglandaceae	abundant	common
hickory, red	*Carya ovalis*	Juglandaceae	abundant	common
hickory, shagbark	*Carya ovata*	Juglandaceae	abundant	no
holly, American	*Ilex opaca*	Aquifoliaceae	occasional in wet woods	no
honeylocust	*Gleditsia triacanthos*	Fabaceae	abundant	common
hophornbeam	*Ostrya virginiana*	Betulaceae	common	no
hoptree, common	*Ptelea trifoliata*	Rutaceae	common along creeks	no
hornbeam, American	*Carpinus caroliniana*	Betulaceae	common along creeks	no

Common name	Species	Family	Bluegrass	Woodland Pastures
kingnut	*Carya laciniosa*	Juglandaceae	common along creeks, limestone uplands	common
leatherwood	*Dirca palustris*	Thymelaeaceae	uncommon	no
locust, black	*Robinia pseudoacacia*	Fabaceae	abundant	common
maple, Amur	*Acer ginnala*	Aceraceae	uncommon, naturalized	no
maple, black	*Acer nigrum*	Aceraceae	abundant	common
maple, hedge	*Acer campestre*	Aceraceae	uncommon, naturalized	no
maple, Norway	*Acer platanoides*	Aceraceae	uncommon, naturalized	no
maple, red	*Acer rubrum*	Aceraceae	common in woods near the Knobs	no
maple, silver	*Acer saccharinum*	Aceraceae	common along rivers	no
maple, sugar	*Acer saccharum*	Aceraceae	abundant	common
mulberry, red	*Morus rubra*	Moraceae	occasional in woods	uncommon
mulberry, white	*Morus alba*	Moraceae	common, naturalized	uncommon, naturalized
ninebark	*Physocarpus opulifolius*	Rosaceae	common on river bluffs, slopes	no
oak, black	*Quercus velutina*	Fagaceae	uncommon	no
oak, bur	*Quercus macrocarpa*	Fagaceae	common	common
oak, cherrybark	*Quercus pagoda*	Fagaceae	widely planted, possibly naturalized	no
oak, chestnut	*Quercus montana*	Fagaceae	occasional in woods near the Knobs, common in Hills of the Bluegrass	no
oak, chinkapin	*Quercus muehlenbergii*	Fagaceae	common	abundant
oak, northern red	*Quercus rubra*	Fagaceae	common	no
oak, overcup	*Quercus lyrata*	Fagaceae	rare, reported in Fayette, Jessamine counties	no
oak, pin	*Quercus palustris*	Fagaceae	common along rivers	no
oak, post	*Quercus stellata*	Fagaceae	common	no
oak, riparian	*Quercus × riparia* (*Q. rubra* × *Q. shumardii*)	Fagaceae	occasional in woods on river slopes	uncommon
oak, scarlet	*Quercus coccinea*	Fagaceae	common in woods near the Knobs	no

oak, shingle	*Quercus imbricaria*	Fagaceae	occasional	uncommon
oak, Shumard	*Quercus shumardii*	Fagaceae	abundant	common
oak, southern red	*Quercus falcata*	Fagaceae	common	no
oak, swamp chestnut	*Quercus michauxii*	Fagaceae	uncommon along rivers	no
oak, swamp white	*Quercus bicolor*	Fagaceae	occasional along Kentucky, Licking rivers, widely planted	no
oak, white	*Quercus alba*	Fagaceae	common, except uncommon in Inner Bluegrass woods	no
oak, willow	*Quercus phellos*	Fagaceae	widely planted, occasionally naturalized in wet areas	no
Osage-orange	*Maclura pomifera*	Moraceae	abundant, naturalized	abundant, naturalized
paper-mulberry	*Broussonetia papyrifera*	Moraceae	common in urban areas, naturalized	no
paulownia, royal	*Paulownia tomentosa*	Paulowniceae	uncommon in disturbed sites, naturalized	no
pawpaw	*Asimina triloba*	Annonaceae	abundant	no
pear, Callery	*Pyrus calleryana*	Rosaceae	common in disturbed urban sites, naturalized	no
pecan	*Carya illinoinensis*	Juglandaceae	reported in Anderson County and along Ohio River	no
persimmon, common	*Diospyros virginiana*	Ebenaceae	common	common
pine, eastern white	*Pinus strobus*	Pinaceae	absent from limestone, common in woods near the Knobs	no
pine, loblolly	*Pinus taeda*	Pinaceae	widely planted, possibly naturalized	no
plum, American	*Prunus americana*	Rosaceae	abundant	occasional
plum, wild goose	*Prunus munsoniana*	Rosaceae	uncommon	no
poplar, white	*Populus alba*	Salicaceae	occasional in disturbed sites, naturalized	no
pricklyash, common	*Zanthoxylum americanum*	Rutaceae	common in woods	uncommon
redbud, eastern	*Cercis canadensis*	Fabaceae	abundant	abundant

Common name	Species	Family	Bluegrass	Woodland Pastures
redcedar, eastern	Juniperus virginiana	Cupressaceae	abundant	abundant
russian-olive	Elaeagnus angustifolia	Elaeagnaceae	naturalized in Jefferson County	no
sassafras	Sassafras albidum	Lauraceae	abundant	occasional
serviceberry, downy	Amelanchier arborea	Rosaceae	abundant	no
silktree	Albizia julibrissin	Fabaceae	naturalized near old home sites	no
sourwood	Oxydendrum arboreum	Ericaceae	in woods near the Knobs, not found on limestone	no
spicebush	Lindera benzoin	Lauraceae	abundant	no
strawberry bush	Euonymus americanus	Celastraceae	common in woods along creeks	no
sugarberry	Celtis laevigata	Ulmaceae	common in western counties along rivers	uncommon
sweetgum	Liquidambar styraciflua	Hamamelidaceae	common along rivers	occasional
sycamore, American	Platanus occidentalis	Platanaceae	abundant	abundant
tree-of-heaven	Ailanthus altissima	Simaroubaceae	common in urban areas, disturbed sites, naturalized	no
walnut, black	Juglans nigra	Juglandaceae	abundant	abundant
willow, black	Salix nigra	Salicaceae	common along rivers	no
willow, Carolina	Salix caroliniana	Salicaceae	uncommon, along Kentucky River	no
willow, crack	Salix fragilis	Salicaceae	common along rivers, naturalized	no
willow, Missouri River	Salix eriocephala	Salicaceae	uncommon	no
willow, sandbar	Salix interior	Salicaceae	common along creeks	no
willow, white	Salix alba	Salicaceae	uncommon along creeks and rivers, naturalized	no
winterberry	Ilex verticillata	Aquifoliaceae	occasional in woods	no
witch-hazel	Hamamelis virginiana	Hamamelidaceae	common in moist woods	no
yellow-poplar, tuliptree	Liriodendron tulipifera	Magnoliaceae	common in woods, but probably not native to the Inner Bluegrass	planted trees are common
yellowwood	Cladrastis kentukea	Fabaceae	common in woods along Kentucky River	no

Notes

Introduction: Two Trees

1. Bodhi tree (*Ficus religiosa*) is also called the sacred fig. Gautama Buddha is said to have achieved enlightenment under a bodhi tree in Bodh Gaya, India (the word *bodhi* can be translated as "enlightenment"). Since that time, bodhi trees have been planted around temples of many faiths throughout tropical Asia. Many of these trees are descendants of the first bodhi tree.

1. The St. Joe Oak: Finding Venerable Trees

1. Andy Mead, "Advocates Want to Save Oak from Getting Axed by Hospital. Tree Is at Root of Dispute," *Lexington (Ky.) Herald-Leader*, June 1, 1990, final ed., sec. City/State.

2. Darren R. Reid, *Daniel Boone and Others on the Kentucky Frontier: Autobiographies and Narratives, 1769–1795* (Jefferson, N.C.: McFarland, 2009).

3. Ursula Davidson, "The Original Vegetation of Lexington, Kentucky, and Vicinity" (M.S. thesis, University of Kentucky, 1950).

4. Mary E. Wharton and Roger W. Barbour, *Bluegrass Land & Life: Land Character, Plants, and Animals of the Inner Bluegrass Region of Kentucky, Past, Present, and Future* (Lexington: University Press of Kentucky, 1991).

5. William S. Bryant et al., "The Blue Ash–Oak Savanna-Woodland, a Remnant of Presettlement Vegetation in the Inner Bluegrass of Kentucky," *Castanea* 45, no. 3 (1980): 149–65.

6. See the appendix for the scientific names of species.

2. The Woodford Groves: The Bluegrass Today

1. See the appendix for species names.

4. Venerable Tree Species

1. Ronald Jones, *Plant Life of Kentucky: An Illustrated Guide to the Vascular Flora* (Lexington: University Press of Kentucky, 2005); Ronald L. Jones and B. Eugene Wofford, *Woody Plants of Kentucky and Tennessee: The Complete Winter Guide to Their Identification and Use* (Lexington: University Press of Kentucky, 2013); B.

Eugene Wofford, *Guide to the Trees, Shrubs, and Woody Vines of Tennessee* (Knoxville: University of Tennessee Press, 2002).

2. Elbert L. Little, *Atlas of United States Trees* (Washington, D.C.: U.S. Department of Agriculture, Forest Service, 1976); A. M. Prasad et al., *A Climate Change Atlas for 134 Forest Tree Species of the Eastern United States* [database], www.nrs.fs.fed.us/atlas/tree (Delaware, Ohio: Northern Research Station, USDA Forest Service, 2007–); U.S. Department of Agriculture, Forest Service, *Forest Inventory and Analysis Program* [database], www.fia.fs.fed.us (Washington, D.C.: USDA, 1930–).

3. Trees are known by many common names, and this is often a source of confusion. Common names in this book follow standard names used in American forestry except for the use of *kingnut* instead of *shellbark hickory*. James W. Hardin, Donald Joseph Leopold, and Fred M. White, *Harlow & Harrar's Textbook of Dendrology*, 9th ed. (Boston: McGraw-Hill, 2000).

4. Only one Latin name is permitted for any organism. The Latin binomial (e.g., *Quercus macrocarpa*) is followed by the name of the author or authors who first described the species. For bur oak, Michx is an abbreviation for André Michaux. The binomial name is followed by the family name, Fagaceae, which tells us that this tree is in the beech family. Latin names can change as a result of research showing changes in our understanding of relationships among species. The Plant Names Project, *The International Plant Names Index* [database], www.ipni.org (London: Royal Botanic Gardens, Harvard University Herbaria, and Australian National Herbarium, 2004–).

5. Sara R. Tanis and Deborah G. McCullough, "Differential Persistence of Blue Ash and White Ash Following Emerald Ash Borer Invasion," *Canadian Journal of Forest Research* 42, no. 8 (2012): 1542–50.

6. F. S. Santamour Jr. and A. J. McArdle, "Checklist of Cultivars of North American Ash (*Fraxinus*) Species," *Journal of Arboriculture* 9, no. 10 (1983): 271–76.

7. George Engelmann, the botanist who gave chinkapin oak its Latin name, apparently misspelled Muhlenberg as Mühlenberg, which was subsequently changed to Muehlenberg. The rules for naming species state that a misspelled word in a validly published name cannot be changed. There are many misspellings in botanical names, but it is against the rules to correct them.

8. For a complete list of trees in the Bluegrass, see the appendix.

9. The fate of black walnut can be followed at www.thousandcankers.com/.

10. Nancy Ross Hugo, "The Mystery of Patrick Henry's Osage-Orange," *American Forests* 109, no. 2 (2003): 32–35; Susan H. Munger, *Common to This Country: Botanical Discoveries of Lewis and Clark* (Sioux City, Iowa: Artisan Books, 2003).

11. Gilbert Imlay, *A Description of the Western Territory of North America* (London: J. Debrett, 1793); Peter Del Tredici, "The Great Catalpa Craze," *Arnoldia* 46, no. 2 (1986): 2–10.

5. The Ingleside Oak: The Bluegrass and the Nashville Basin in 1779

1. Mary E. Wharton and Roger W. Barbour, *Bluegrass Land & Life: Land Character, Plants, and Animals of the Inner Bluegrass Region of Kentucky, Past, Present, and Future* (Lexington: University Press of Kentucky, 1991).

2. E. Lucy Braun, "The Phytogeography of Unglaciated Eastern United States

and Its Interpretation," *Botanical Review* 21, no. 6 (1955): 297–375; Braun, *Deciduous Forests of Eastern North America* (1950; repr., Caldwell, N.J.: Blackburn Press, 2001).

3. Hanson is quoted in Wharton and Barbour, *Bluegrass Land & Life*, 23; James Nourse, "Journey to Kentucky in 1775. Diary of James Nourse," *Journal of American History* 19, nos. 2–4 (1925): 121–28, 251–60, 351–64.

4. Giant cane, often just called cane in the Bluegrass and Nashville Basin, is *Arundinaria gigantea*, also called river cane and switch cane. Two other species, hill cane (*A. appalachiana*) and switch cane (*A. tecta*), are found farther south and east.

5. On the basis of modern phylogenetic analysis, all cattle, including American and European bison, auroch, yak, and others, should be considered as one genus, *Bos*. Understanding these relationships is critical to our discussion of conservation, and especially to our understanding of how our woodland pastures compare to those of Europe. Dorian Garrick and Anatoly Ruvinsky, eds., *The Genetics of Cattle*, 2nd ed. (Boston: CAB International, 2014).

6. Filson's accounts are not always reliable because his goal was to promote the settlement of Kentucky. There are many similar stories from other travelers and historians that corroborate his observations. John Filson, *The Discovery, Settlement and Present State of Kentucke* (1784; repr., Westminster, Md.: Heritage Books, 2007), 32–33.

7. James Gettys McGready Ramsey, *The Annals of Tennessee, to the End of the Eighteenth Century* (Philadelphia: Lippincott, Grambo, 1853).

8. The ancient bison (*Bison antiquus*) has been extinct for about 10,000 years; the long-horned bison (*Bison latifrons*), found near Big Bone Lick, has been extinct for about 20,000 years.

9. This description is based on Joel Asaph Allen, *History of the American Bison: Bison Americanus* (Washington, D.C.: U.S. Government Printing Office, 1877); William Temple Hornaday, *The Extermination of the American Bison* (1887; repr., Washington, D.C.: Smithsonian Institution Press, 2002); Andrew C. Isenberg, *The Destruction of the Bison: An Environmental History, 1750–1920* (New York: Cambridge University Press, 2001); Dale F. Lott, *American Bison: A Natural History* (Berkeley: University of California Press, 2002); Ernest Thompson Seton, *The American Bison or Buffalo* (New York: Charles Scribner's Sons, 1906).

10. Our knowledge of ancient woodland pastures in Europe depends on the work of a number of European scientists who have devoted their careers to understanding the role of ancient woodland pastures. Key publications are Tibor Hartel et al., "Wood-Pastures in a Traditional Rural Region of Eastern Europe: Characteristics, Management and Status," *Biological Conservation* 166 (October 2013): 267–75; Oliver Rackham, *Ancient Woodland: Its History, Vegetation and Uses in England* (London: E. Arnold, 1980); Ian D. Rotherham, ed., *Trees, Forested Landscapes and Grazing Animals: A European Perspective on Woodlands and Grazed Treescapes* (London: Routledge, 2013); F. W. M. Vera, *Grazing Ecology and Forest History* (New York: CABI Publications, 2000); Tibor Hartel and Tobias Plieninger, *European Wood-Pastures in Transition: A Social-Ecological Approach* (London: Routledge. 2014).

11. Darren R. Reid, *Daniel Boone and Others on the Kentucky Frontier: Autobiographies and Narratives, 1769–1795* (Jefferson, N.C.: McFarland. 2009). In another narrative, Collins claimed a substantial number of fights with Indians, but at the

time he was trying to obtain a generous pension on the basis of his military experience. The earlier narrative is more consistent with accounts by other early settlers.

12. Lucien Beckner, "John Findley: The First Pathfinder of Kentucky," *Filson Club History Quarterly* 1, no. 3 (1927): 114; Lucien Beckner, "Eskippakithiki: The Last Indian Town in Kentucky," *Filson Club History Quarterly* 5 (1932): 1–328.

13. Martin Wymore is quoted by Wharton and Barbour, *Bluegrass Land & Life*, 36, and more fully in Draper Manuscript, Kentucky Papers, CC 12:92 Wisconsin Historical Society, Madison.

14. This remarkable ability to study past drought is the product of many decades of work by dendrochronologists, scientists who use tree rings to analyze past climate and other events, which culminated in a series of publications by Edward R. Cook and his colleagues at Columbia University's Lamont-Doherty Earth Observatory in Palisades, N.Y. Edward R. Cook et al., "North American Drought: Reconstructions, Causes, and Consequences," *Earth-Science Reviews* 81, no. 1 (2007): 93–134. Neil Pederson et al., "The Legacy of Episodic Climatic Events in Shaping Temperate, Broadleaf Forests," *Ecological Monographs* 84 (April 2014): 599–620; Ryan W. McEwan and Brian C. McCarthy, "Anthropogenic Disturbance and the Formation of Oak Savanna in Central Kentucky, USA," *Journal of Biogeography* 35, no. 5 (2008): 965–75.

15. Cook et al., "North American Drought"; Pederson et al., "The Legacy of Episodic Climatic Events"; McEwan and McCarthy, "Anthropogenic Disturbance and the Formation of Oak Savanna."

6. The Woodland Pasture

1. For a detailed analysis of European wood pastures, see Tibor Hartel and Tobias Plieninger, eds., *European Wood-Pastures in Transition: A Social-Ecological Approach* (New York: Routledge, 2014).

2. Oliver Rackham, *Woodlands,* 2nd ed. (London: Collins, 2012).

3. Phylogenetic analyses treat the auroch either as a separate species, *Bos primigenius*, or as a subspecies of modern cattle, *Bos taurus primigenius*. It is clear that aurochs were grazing animals like cattle, not browsers.

4. The European moose is called elk in Europe, whereas in North America *elk* is a term reserved for the wapiti (*Cervus canadensis*). The European and American moose (*Alces americanus*) are similar species with overlapping ranges in Siberia, and they are called European elk and Siberian elk, respectively, in Europe.

5. Red deer is closely related to the American elk or wapiti (*Cervus canadensis*), not to the much smaller white-tailed deer (*Odocoileus virginianus*).

6. The wild European horse was probably the tarpan. Smaller than the modern domesticated horse, it is probably the ancestor—or one of the ancestors—of today's horse.

7. F. W. M. Vera, *Grazing Ecology and Forest History* (New York: CABI, 2002).

8. Hervé Bocherens et al., "European Bison as a Refugee Species? Evidence from Isotopic Data on Early Holocene Bison and Other Large Herbivores in Northern Europe," *PLoS ONE* 10, no. 2 (2015): e0115090.

9. Roger C. Anderson, James S. Fralish, and Jerry M. Baskin, *Savannas, Barrens, and Rock Outcrop Plant Communities of North America* (New York: Cambridge University Press, 1999).

10. The prolonged drought period that began in 625 was well before the well-known Medieval Climate Anomaly (1400–1700), which did not create a significant drought signature in this region. Michael E. Mann et al., "Global Signatures and Dynamical Origins of the Little Ice Age and Medieval Climate Anomaly," *Science* 326, no. 5957 (2009): 1256–60.

11. Neil Pederson et al., "The Legacy of Episodic Climatic Events in Shaping Temperate, Broadleaf Forests," *Ecological Monographs* 84 (April 2014): 599–620.

12. Alan L. Olmstead and Paul W. Rhode, "Reshaping the Landscape: The Impact and Diffusion of the Tractor in American Agriculture, 1910–1960," *Journal of Economic History* 61, no. 3 (2001): 663–98.

7. The Mother Tree: Reproduction in Venerable Trees

1. The science journalist Sharon Levy has written a fine, nontechnical book about these giant animals of North America: Sharon Levy, *Once and Future Giants: What Ice Age Extinctions Tell Us about the Fate of Earth's Largest Animals* (New York: Oxford University Press, 2011).

2. Like many ideas in biology, the concept of mast fruiting as an evolutionary strategy originated with Charles Darwin. Daniel Janzen has formalized our understanding of mast fruiting: Daniel H. Janzen, "Seed Predation by Animals," *Annual Review of Ecology and Systematics* 2 (1971): 465–92; Janzen, "Why Bamboos Wait So Long to Flower," *Annual Review of Ecology and Systematics* 7, no. 1 (1976): 347–91.

8. The Guardians: Trees in Cemeteries

1. The U.S. champion American basswood in Lexington Cemetery is 86 inches in diameter and 96 feet tall. The American smoketree is 32 inches in diameter with multiple stems, and 42 feet tall. In 2012 a slightly larger smoketree in Connecticut replaced this tree as the national champion. The champion tree database is maintained by the American Forests organization and is available online at www.americanforests.org/our-programs/bigtree/.

2. Information about the Lexington Cemetery is from interviews with the cemetery's horticulturist, Miles Penn, and from Burton Milward and Lexington Cemetery Company, *A History of the Lexington Cemetery* (Lexington: Lexington Cemetery Co., 1989).

3. This discussion of the evolution of cemeteries and parks is based on Milward and Lexington Cemetery Co., *A History of the Lexington Cemetery*, and on David Charles Sloane, *The Last Great Necessity: Cemeteries in American History* (Baltimore: Johns Hopkins University Press, 1991).

9. The Loudon Grove: Trees in Public Spaces

1. Galen Cranz, *The Politics of Park Design: A History of Urban Parks in America* (Cambridge: MIT Press, 1982); Sara Cedar Miller, *Central Park: An American Masterpiece* (New York: H. N. Abrams, 2003); Friends of the Public Garden, *Boston Common* (Charleston, S.C.: Arcadia Publishing, 2005).

11. The Elmwood Trees: Growing Old Gracefully

1. There has been very little scientific study of the effects of lightning strikes on trees. The observations in this chapter are mostly my own, supplemented with published literature. Most of the published literature is anecdotal; there is only one paper that attempts a quantitative analysis: Jakke Mäkelä et al., "Attachment of Natural Lightning Flashes to Trees: Preliminary Statistical Characteristics," *Journal of Lightning Research* 1 (2009): 9–21.

12. The Floracliff Trees: The Long Lives of Venerable Trees

1. William S. Bryant et al., "The Blue Ash–Oak Savanna-Woodland, a Remnant of Presettlement Vegetation in the Inner Bluegrass of Kentucky," *Castanea* 45, no. 3 (1980): 149–65. These are estimated ages in 1980 from counts of tree rings. The estimates are probably not as accurate as those of Neil Pederson et al., "The Legacy of Episodic Climatic Events in Shaping Temperate, Broadleaf Forests," *Ecological Monographs* 84 (April 2014): 599–620, or of Ryan W. McEwan and Brian C. McCarthy, "Anthropogenic Disturbance and the Formation of Oak Savanna in Central Kentucky, USA," *Journal of Biogeography* 35, no. 5 (2008): 965–75.

13. The Old Schoolhouse Oak: Extending the Lives of Old Trees

1. *Lexington Herald-Leader,* August 8, August 29, October 14, October 27, and October 29, 2008; October 5, October 8, August 3, 2014; and letters to the editor, October 8, November 19, December 18, 2013.

14. The Future of Venerable Trees and Woodland Pastures

1. Plants reproduced by cuttings result in ramets—all the individuals of the original tree that are genetically identical. The process of creating ramets is called cloning, and the individuals are often called clones. Clones that occur naturally in a single place, such as a stand of black locust arising from a single root system, are called genets. All "October Glory" red maples, for example, are genetically identical ramets, clones of the original tree from which cuttings were taken. In 1970 the Plant Variety Protection Act extended intellectual property rights to some plants reproduced by seeds, though this does not apply to any trees. The International Union for the Protection of New Varieties of Plants extends intellectual property rights to plants internationally. Office of Technology Assessment, U.S. Congress, *New Developments in Biotechnology: Patenting Life* (New York: Dekker, 1990); Joe Miller, "Patent Law: How Patents Grew over Time to Include Living Organisms," *Cooking Up a Story,* http://cookingupastory.com/patent-law-how-patents-grew-over-time-to-include-living-organisms.

2. This is something of a moot point now, since most white ash trees are being killed by the emerald ash borer.

3. The history of breeding forest trees is the exact opposite of horticultural selection. Foresters breed trees to increase genetic diversity because it is too risky to plant a narrow genetic base of a long-lived species in a plantation. In that sense, tree breeding for forestry resembles the breeding of animals more than the

breeding of crop plants. Forest tree breeding is focused on long-term performance, whereas horticultural selection and breeding are focused on short-term performance. Timothy L. White, W. T. Adams, and David B. Neale, *Forest Genetics* (Cambridge, Mass.: CABI, 2007).

4. Jerry M. Melillo, T. C. Richmond, and Gary W. Yohe, *Climate Change Impacts in the United States: The Third National Climate Assessment* (Washington, D.C.: U.S. Global Change Research Program), http://nca2014.globalchange.gov/report. James Bruggers has created an excellent introduction to the effects of climate change on Kentucky, "Climate Change Warming Kentucky, Indiana, Report Says," *Louisville Courier-Journal,* May 7, 2014, www.courier-journal.com/story/tech/science/environment/2014/05/06/national-climate-assessment-southeast-region/8765423/.

5. Baek-Min Kim et al., "Weakening of the Stratospheric Polar Vortex by Arctic Sea-Ice Loss," *Nature Communications* 5, no. 4646 (September 2014).

6. Louis R. Iverson et al., *Atlas of Current and Potential Future Distributions of Common Trees of the Eastern United States* (Radnor, Pa.: U.S. Department of Agriculture, Forest Service, Northeastern Research Station, 1999); Louis R. Iverson and Anantha M. Prasad, "Potential Redistribution of Tree Species Habitat under Five Climate Change Scenarios in the Eastern US," *Forest Ecology and Management* 155, no. 1 (2002): 205–22.

Index

Unless otherwise noted, place names are in Kentucky. Italic page numbers refer to illustrations.

Algonquian, Virginia (language), 59, 78, 155
Algonquin people, 59
ash: blue (*Fraxinus quadrangulata*), 2–3, 5, 10, 12–14, 28, 30, 37, 41, 43, 45, 52–55, 76, 82, 84, 96, 100, 102, 105, 106, 113, 116, 118–20, 122, 123, 127, 131, 134, 137, 141, 154–56, 162, 163, 165, 167, 171–73, 183, 184, 187, 195, 203, 212, 222n1; green (*Fraxinus pennsylvanica*), 55, 193, 212; white (*Fraxinus americana*), 10, 23, 24, 54–55, 67, 105, 116, 133, 136, 193, 203, 212, 222n2
auroch (*Bos primigenius*), 87, 90, 219n5, 220n3

Ball Homes, 177
Barbour, Roger, 71
basswood, American (*Tilia americana*), 113–14, 118–19, 212, 221n1
Beaumont Farm, 134, 136
beech, American (*Fagus grandifolia*), 2, 212
Belle Meade, Nashville, Tenn., 43
Bell family, 152
Big Beaver Tree Service, 169, 180
bison: American (*Bos bison bison*), 20, 74, 75–81, 84, 85, 90, 92–96, 105–6, 107, 110, 151, 152, 158, 195, 197, 219n5; European (*see* wisent)
Blackford Oaks neighborhood, 145, 146
blockhouse, 9, 10, 71
Bluegrass (region): Inner, 7, 8, 17, 19–23, 26, 27, 31, 65, 67, 69, 78, 80, 108, 137, 157; Outer, 19–23, 27, 78, 92, 104
bluegrass, Kentucky (*Poa pratensis*), 17, 19
Bluegrass Conservancy, 23
bodark. *See* Osage-orange
bodhi (*Ficus religiosa*), x, 1, 217n1
Boswell Woods (Fayette County), 118
Boyle County, Ky., 137
Braun, E. Lucy, 73, 84, 95
buckeye: Ohio (*Aesculus glabra*), 27, 66, 130, 212; yellow (*Aesculus flava*), 27, 212
buffalo. *See* bison
buffalo trace, 71, 94, 141, 156, 177
Burbank, Luther, 192

Calumet Farm (Fayette County), 155
cambium, vascular, 160, 170
cane, giant (*Arundinaria gigantea*), 15, 17, 20, 31, 50, 69, 73, 74–75, 85, 91–96, 107, 110, 139–40, 169, 196, 197, 219n4
Cane Run (Fayette County), 139
Castlewood Park. *See* Loudon House/ Loudon Grove

225

catalpa, northern (*Catalpa speciosa*), 70, 212
Catholic Diocese of Lexington, 130
cattle (*Bos taurus*), 10, 12, 15, 17, 27, 35, 36, 67, 71, 77, 90, 95, 104, 106, 110, 116, 123, 139, 151–53, 155, 157, 194–97, 219n5, 220n3
cedar. *See* redcedar, eastern
cedar glade, 35, 37, 41, 54
cemetery, rural, 117–18
Centre College (Boyle County), 133
Chandler, Toss, 106, 195
Chandler Farm (Woodford County), 106, 110, 111
Cherokee people, 78, 80–81
cherry, black (*Prunus serotina*), 68, 73, 212
Cincinnati Arch, 14
Clark's Run (Boyle County), 137
Clay, Henry, 114, 116, 123
climate change, 45, 198–208
clones, 67, 70, 187, 188, 192, 194, 195, 222n1
clover, running buffalo (*Trifolium stoloniferum*), 197
coffeetree, Kentucky (*Gymnocladus dioicus*), 67, 212
Coldstream Farm (Fayette County), 139–42
College Grove, Tenn., 35
Collins, Josiah, 9–10, 50, 71, 74, 75, 78, 197, 219n11
Cumberland River, 44, 56, 58, 59, 73, 80

Danville (Boyle County), 126, 132, 133, 137, 183
Davidson, Ursula, 12–14
decay, 135, 153, 155, 157, 160–62, 164, 167, 170–71, 173, 175, 183, 186, 196
deer: red (*Cervus elaphus*), 78, 90, 220n5; white-tailed (*Odocoileus virginianus*), 77, 104, 220n5
dendrochronology, 82
detention basin, 107, 108, 142

Dixie Elementary School (Fayette County), 126–27
Downing, Andrew Jackson, 116, 118
drought, 20, 22, 23, 45, 47, 49–50, 59, 64, 79–81, 82–84, 92, 94–96, 104, 109–11, 130, 167, 194, 196–97, 198, 203, 220n14, 221n10

Eden Shales. *See* Hills of the Bluegrass
elk: American (*Cervus canadensis*), 77, 104, 220n4; red (*see* deer: red)
Elkhorn Creek (Fayette County), 30, 73, 92–94
Ellwanger and Barry Nursery, New York, 191
elm: American (*Ulmus americana*), 10, 67, 73, 105, 167, 187, 213; rock (*Ulmus thomasii*), 67, 213; slippery (*Ulmus rubra*), 67, 213
Elmwood Stock Farm (Scott County), 151–58, 194
emerald ash borer (*Agrilus planipennis*), 54–55, 67, 136, 187, 222n2
Eskippakithiki (Clark County), 78–81, 91

Farmall tractor, International Harvester, 96–97
Farrar, Asa, 10
Fayette Alliance, 23
Fayette County, Ky., 12–13, 18, 23, 26, 70, 150
fescue (*Festuca* spp.), 33, 197
Filson, John, 73, 76, 93, 94, 219n6
fire, 49, 64, 74, 81–82, 87, 92, 194
Floracliff State Nature Preserve (Fayette County), 165
forage, 15, 19, 77, 94, 95, 194
Forest Inventory and Analysis Program, 48
Fort Ancient (culture), 81
Fort Boonesborough (Madison County), 10, 69, 74
Fort Harrod. *See* Harrodsburg (Mercer County)
Franklin County, Ky., 106

genet, 66, 222n1
grazers/grazing, 15, 30, 33, 71, 77–78, 87, 90–91, 94–96, 104–6, 110–11, 121, 151–53, 155, 175, 194–97, 220n3
Griffin Gate Marriott Resort and Spa (Fayette County), 141–42, *172*
Griffith Woods Wildlife Management Area (Harrison County), 92, 99, 104–6, 110, 194, 196, 197
groundhog. *See* woodchuck
ground sloth, giant, 104

hackberry (*Celtis occidentalis*), 47, 65, 73, 213
Hamburg Giant Grove (Fayette County), *148*, 150
Hanson, Thomas, 73
Harrodsburg (Mercer County), 9, 10, 78, 128, 146, 148
Hartel, Tibor, 90
hawthorn (*Crataegus* spp.), 78, 213
Headley, Hal Price, 134, 136
hedge-apple. *See* Osage-orange
hickory: bitternut (*Carya cordiformis*), 62, 66, 213; mockernut (*Carya tomentosa*), 62, 66, 213; pignut (*Carya glabra*), 62, 66, 213; red (*Carya ovalis*), 62, 66, 213; shagbark (*Carya ovata*), 37, 41, 62, 64, 65, 66, 213; shellbark *(see* kingnut)
Hills of the Bluegrass, 19, 21, 22, 26, 68, 92
Hoffman, Ian, 169, 180
honeylocust (*Gleditsia triacanthos*), 66, 78, 105, 213
honeysuckle, bush (*Lonicera* spp.), 31, 33, 196, 197
horse (*Equus caballus*), 2, 12, 16, 19, 71, 90, 95, 151, 152, 155, 194, 195, 196, 197, 220n6
Huskisson Farm (Woodford County), 106, 110
hybrid trees, 27, 28, 30, 54, 58, 61, 65, 67

Ingleside Oak (Fayette County), 71–72, 142–44, *177*

Inner Bluegrass. *See* Bluegrass (region): Inner
Inner Nashville Basin. *See* Nashville Basin: Inner

James, Beverly, 165
Julian, Jane/Julian Farm, 106, 110, 194, 197

karst, 20, 22–24, 39, 59, 80, 82, 84, 92, 94, 96, 194
Kentucky bluegrass (plant). *See* bluegrass, Kentucky
Kentucky River, 27–30, 44, 54, 58, 69, 78, 106, 165
Kentucky State Nature Preserves Commission, 30
kingnut (*Carya laciniosa*), 2, 5, 10, 13, 14, 28, 30, 43, 45, 61–65, 84, 99–111, 113, 118–20, 127, 136, 141, 167, 187, 195, 196, 203, 214, 218n3
Kissing Tree, Transylvania University, 133
Knight, Thomas, 72, 91, 142
Knobs Region, 20, 78, 80

Leestown (Franklin County), 30, 93
Legacy Trail (Fayette County), 139–41
Lewis, Meriwether, 69
Lexington Cemetery (Fayette County), 13, 50, 113–20
Licking River, 28, 30, 56, 59, 92
lightning, 2, 53, 82, 153, 154, 158–64, 170, 171, 182, 222n1
limestone, 14, 17, 19, 20, 22, 23, 24, 26–28, 35, 37, 39, 43, 47, 50, 53, 59, 64, 66, 67, 69, 82
Limestone, Lexington, 19, 23
Lipscomb Academy Elementary School, Nashville, Tenn., 127
Little, Elbert, 48
locust, black (*Robinia pseudoacacia*), 66, 78, 222n1
Loudon House/Loudon Grove, 121–25

maple: black (*Acer nigrum*), 10, 47, 68,
    199, 203; red (*Acer rubrum*), 138,
    187, 189, 193, 222n1; silver (*Acer
    saccharinum*), 214; sugar (*Acer
    saccharum*), 10, 47, 68, 95, 118, 199,
    201, 203
Masterson Station Park (Fayette
    County), 123–25
mast fruiting, 58, 64, 109, 221n2
McCarthy, Brian, 84–85, 167
McEwan, Ryan, 84–85, 167
McGrathiana Farm. *See* Coldstream
    Farm
meranti, red (*Shorea macroptera*), 158
meristem, apical, 170
migration, assisted, 203
moose, European (*Alces alces*), 90,
    220n4
mossycup oak. *See* oak: bur
mother tree (Harrison County), 99–111
Mount Auburn Cemetery,
    Massachusetts, 117–18, 121
Mt. Olivet Cemetery, Nashville, Tenn.,
    118
mulberry: red (*Morus rubra*), 70, 73,
    214; white (*Morus alba*), 70, 214
Murfreesboro, Tenn., 38, 73

Nashville Basin, 2, 7, 10, 14, 15, 16, 20,
    23, 35–45; Inner, 7, 8, 35–37, 54, 73,
    96; Outer, 37–38, 73
Nashville City Cemetery, Nashville,
    Tenn., 118, 105
Natchez Trace, 43
Nature Conservancy, 30
Nourse, James, 73

oak: black (*Quercus velutina*), 56; bur
    (*Quercus macrocarpa*), 1–16, 35–36,
    42, 45, 48–51, 71–73, 104, 106, 107,
    113, 115, 118–20, 123–30, 133–35, 138,
    140–41, 143–44, 146, 150, 153, 159, 161,
    165, 167, 169, 174, 177–83, 185, 187,
    195, 197, 199, 203, 204, 214, 218n4;
    chinkapin (*Quercus muehlenbergii*),
    2, 5, 10, 13, 14, 19, 28, 30, 41, 43, 50,
    58–61, 84, 96, 105, 106, 113, 118–20,
    123, 127, 134, 137, 138, 141, 165, 167,
    187, 195, 203, 208, 214, 218n7; Deam
    (*Quercus muehlenbergii* × *alba*),
    61; northern red (*Quercus rubra*),
    28, 30, 56, 58, 191; pin (*Quercus
    palustris*), 14, 28, 56, 58; riparian
    (*Quercus rubra* × *shumardii*), 30;
    shingle (*Quercus imbricaria*), 28,
    56; Shumard (*Quercus shumardii*),
    5, 10, 13, 14, 28, 30, 33, 37, 41, 43,
    55–58, 84, 113, 117–20, 138, 141, 147,
    167, 181, 187, 195, 203, 206, 215;
    southern red (*Quercus falcata*), 28,
    56; white (*Quercus alba*), 41, 43, 47,
    61
Oak Hill neighborhood (Woodford
    County), 148–50
Old Schoolhouse Oak (Fayette County),
    177–85
Olmsted, Frederick Law, 116
Ordovician Period, 19, 37
Osage-orange (*Maclura pomifera*),
    69–70, 126, 215
Outer Basin. *See* Nashville Basin: Outer
Outer Bluegrass. *See* Bluegrass (region):
    Outer

Palisades, Kentucky River, 27, 30, 69
parenchyma cells, 170
Patterson, Robert, 9, 10
Patterson Cabin (Fayette County), 10, *11*
Pederson, Neil, 95, 165, 167
phloem, 160, 170
phosphorus, 19, 35, 37
Plant Patent Act of 1930, 192
plum, American (*Prunus americana*),
    78, 215
pollen/pollination, 100–102, 109
poplar. *See* yellow-poplar
preservation, 177–78, 180–81, 183, 186,
    197

Rackham, Oliver, 87, 90, 219n10
Radnor Lake State Natural Area, Nashville, Tenn., 40, 42, 43
ramet, 187, 222n1
redcedar, eastern (*Juniperus virginiana*), 20, 28, 37, 68–69, 92, 216
Reforest the Bluegrass, 31–33
regeneration, natural, 33, 50, 60, 90, 96, 97, 111, 158, 194, 195, 196
rivercane. *See* cane, giant
RR Donnelley (Boyle County), 137
rural cemetery, 117–18

sacred fig. *See* bodhi
Salt River, Kentucky, 28, 30, 56, 59, 92
savanna, 49, 64, 73, 81, 82, 91–92
shade tolerance, 15, 48, 54, 58, 60, 64, 69, 74, 94
Shawnee people, 64, 78–81
sheep, domestic (*Ovis aries*), 71, 95, 151–53
smoketree, American (*Cotinus obovatus*), 113, 221n1
Squire Oak neighborhood (Fayette County), 142
St. Joe Oak (Fayette County), 5–16
Stark, Paul, 192
Stark Bros. Nursery, Missouri, 192
stem cells, 170
Stone, Mac, 152
sugarberry (*Celtis laevigata*), 47, 68, 216
Sullivan University (Fayette County), 134, 136
sycamore, American (*Platanus occidentalis*), 69, 216

Town Branch Creek (Fayette County), 30
trace buds, 170
Transylvania University (Fayette County), 10, 11, 133, 207
tree pens, 106, 111, 158, 195–96, 197

tulip-poplar or tulip tree. *See* yellow-poplar

University of Kentucky (Fayette County), 12, 13, 133–34, 138, 139
urban forest, urbanization, 1, 14, 23

Vera, Frans, 87, 90
Veteran's Oak (Fayette County), 125, 126

walnut, black (*Juglans nigra*), 10, 11, 14, 65–66, 73, 202, 203, 216
wapiti. *See* elk: American
Warner Parks, Nashville, Tenn., 40, 42–43
Westphal, Donna, 7
Wharton, Mary, 13, 71
Wilderness Road, Kentucky, 10
Willman Way neighborhood (Fayette County), 107, 110
Wilmore Camp Meeting Ground (Jessamine County), 132
winter creeper (*Euonymus fortunei*), 33
wisent (*Bos bison bonasus*), 77–78, 85, 87, 90
woodchuck (*Marmota monax*), 155
Woodford County, Ky., 17–18, 21, 27, 106, 148–50, 157
Woodland Park (Fayette County), 123–24
woodland pasture, 5, 7, 10, 12, 14, 15, 17–19, 21, 23, 25, 27, 28, 30, 33, 35, 37, 40, 41, 43, 45, 47, 50, 53–70, 71, 74, 77, 84–85, 87–97, 104–7, 110–11, 113, 116, 118, 123, 126, 127, 130, 132, 133–34, 136–39, 141–42, 144, 146, 148, 150, 151–58, 168, 175, 189–99, 203

xylem, 160, 170

yellow-poplar (*Liriodendron tulipifera*), 68, 216
yellowwood (*Cladrastis kentukea*), 28, 216